Natural Hazards Analysis

Reducing the Impact of Disasters

Natural Hazards Analysis

Reducing the Impact of Disasters

John C. Pine

CRC Press
Taylor & Francis Group
Boca Raton London New York

CRC Press is an imprint of the
Taylor & Francis Group, an **informa** business

Auerbach Publications
Taylor & Francis Group
6000 Broken Sound Parkway NW, Suite 300
Boca Raton, FL 33487-2742

Printed in the United States of America on acid-free paper
10 9 8 7 6 5 4 3 2 1

International Standard Book Number-13: 978-1-4200-7038-5 (Hardcover)

Library of Congress Cataloging-in-Publication Data

Pine, John C., 1946-
Natural hazards analysis : reducing the impact of disasters / John Pine.
p. cm.
Includes bibliographical references and index.
ISBN 978-1-4200-7038-5
1. Disasters--Social aspects. 2. Emergency management. 3. Preparedness. 4. Community organization. 5. Disasters—Risk assessment. 6. Risk management. I. Title.

HV553.P528 2008
363.34'7--dc22 2008023245

Visit the Taylor & Francis Web site at
http://www.taylorandfrancis.com

and the Auerbach Web site at
http://www.auerbach-publications.com

Contents

Preface

Damage and economic impacts of natural and man-made disasters have been growing at exponential rates over the past few decades. The magnitude of these disasters and our global interconnectedness have grown to the point where disasters may have regional, national, and even global economic consequences. We have attempted to reduce risks to human, economic, and environmental infrastructure, but at the same time continue to develop vulnerable geographic coastline areas of the world. Our communities are more vulnerable today than ever before to natural and human-caused disasters. Communities that are subject to such high degrees of risk can only achieve sustainable social and economic development through appropriately informed planning, protection, mitigation, and recovery strategies that are based on research that integrates our complex social, economic, and environmental systems.

Natural Hazards Analysis: Reducing the Impacts of Disasters examines the hazards analysis process but also provides a broad view of how we use the outcomes in preparedness, response, recovery, and mitigation activities of the emergency management process. Specifically, this book demonstrates how we utilize a hazards analysis for developing hazard mitigation strategies and the broader hazards risk management process. Further, the roles of problem solving, decision making, and risk communication are stressed in this text as a means of ensuring that we accurately identify problems associated with hazards and the risks that they present and make quality decisions associated with these risks. The risk communication process is also addressed, and the importance of engaging stakeholders in ensuring that risks are understood is stressed. The key is that hazards analysis is not an isolated process, but one that involves public officials and employees, the public, and the media, and that care is taken to ensure that all appreciate the risks that confront our organizations and communities.

Social and natural scientists currently apply various technological tools to characterize hazards, but we still have a limited understanding of the social, economic, cultural, and environmental impacts of hurricanes, floods, earthquakes, chemical spills, or terrorist incidents. The fundamental processes that the natural, social, and engineering sciences have to consider include a very dynamic landscape, which requires policies that promote adaptation and resiliency rather than a philosophy of control. Policies and decisions associated with rebuilding after Hurricanes Katrina and Rita demonstrate the difficulty in understanding risk and vulnerability in changing natural, social, and built environments.

The key to understanding the dynamics of risk and vulnerability to disasters is through a comprehensive hazards analysis. A cross-disciplinary approach is recommended in this text to clarify how to deal with uncertainty in ever-changing social, economic, and natural landscapes in disasters. The intent of this text is to demonstrate how we use outputs from environmental models, the tools of geographic information systems (GIS), spatial analysis, and remote sensing to clarify the relationships between human, economic, and natural systems.

We stress that a better understanding of our vulnerabilities will facilitate community planning, hazard mitigation, and our ongoing rebuilding, restoration, and recovery efforts. Few books take the broad-based approach, as recommended in *Natural Hazards Analysis: Reducing the Impacts of Disasters,* in examining the impacts of disasters on our natural, social, cultural, and economic environments. This broader view will help us to build sustainable communities.

The goals of the book are:

1. To provide a framework for understanding the nature and consequences of natural and human-caused hazards on the natural, built, and human environments
2. To examine strategies that may be taken at the individual, organizational, community, or regional levels to reduce the adverse consequences of disasters and foster sustainability

Primary Learning Objectives

- To be able to explain the role and uses of hazards analysis in organizational decision making and public policy
- To be able to characterize the nature of hazards and the factors that affect them
- To be able to explain the nature and types of models used to characterize natural and technological hazards
- To be able to explain how we measure potential and actual direct and indirect social, economic, and environmental impacts of disasters on families, businesses, and communities
- To be able to describe the role and uses of spatial analysis and geographic information systems in hazards analysis
- To be able to clarify how individuals perceive risk and how risk perception influences decision making at the individual, business, and community level
- To be able to explain how risk management and hazard mitigation strategies could reduce or minimize losses from disasters
- To be able to explain the role of hazards analysis in building and fostering sustainable communities and business organizations

Acknowledgments

This text is an outgrowth of my teaching and research efforts over the past ten years at Louisiana State University (LSU). I began teaching Environmental Hazards Analysis for both graduate and undergraduate students at LSU in 1996 and used in this class my experiences from conducting hazards analysis projects for local, state, and federal agencies as well as private-sector chemical processing operations. To say the least, I would not have been able to initiate this text without the support from the Federal Emergency Management Agency (FEMA), the U.S. Environmental Protection Agency (EPA), the National Oceanographic Atmospheric Administration (NOAA), the National Science Foundation, the Organization of American States, Louisiana Office of Emergency Management (LA OEP), the Louisiana State Police, the Louisiana Department of Environmental Quality (DEQ), and numerous state and local government agencies throughout the United States. I want to thank Ms. Mary Lee Eggart, who is the graphic artist in the Department of Geography and Anthropology at LSU for her creative contribution to the graphics in this text. I especially want to acknowledge and thank my colleagues Gavin Smith at the University of North Carolina and Greg Shaw at George Washington University (GWU) for their contributions in Chapters 8 through 10. Gavin saw first hand how a quality hazards analysis is critical in community planning during recovery efforts following Hurricane Katrina. His ability to clearly articulate hazard mitigation initiatives is based on a sound understanding of hazards and decision making. Greg has taught classes at GWU and developed several courses for FEMA's Higher Education Project associated with hazards risk management. His insights into the nature of organizational risk are seen throughout this text, as well as his view that hazards risk management must fit within an organization's short- and long-term strategies. We thus see that hazards analysis is part of community and organizational decision making as it relates to understanding risk, especially from hazards, and that organizational and community policies are critical in reducing our vulnerabilities to the adverse impacts from disasters.

I would like to also acknowledge my social science colleagues at LSU, the University of New Orleans (UNO), and throughout the United States in reorienting my view of the interactions between social, economic, and environmental systems. For many years, I examined how human actions impacted our physical landscapes. As a result of my work with Shirley Laska and Kristina Peterson at UNO, and Breda Phillips at Oklahoma State University, and families in the Grand Bayou community located twenty-five miles south of the City of New Orleans, I became more aware of how hazards associated with natural systems and the environment impact on individuals, families, and communities. Community members were engaged in our hazards analysis process from day one and participated jointly with us in expanding our understanding of vulnerability and communicating our findings at numerous professional meetings throughout the United States. Community engagement and open risk communication are stressed in this text and included as a vital part of the hazards analysis process.

John C. Pine
Director, Disaster Science & Management
Louisiana State University

Chapter 1

Introduction to Hazards Analysis

Objectives

The study of this chapter will enable you to:

1. Clarify why hazards analysis is so important in reducing losses from disasters.
2. Compare and contrast hazards terminology.
3. Explore the view of extreme hazard events as a primary cause of disasters.
4. Explain alternative hazard paradigms that include social, political, economic, and environmental systems.
5. Define the hazards analysis process and its links to hazards risk management and comprehensive emergency management.
6. Explain why communicating risk is so critical in a hazards analysis.

Key Terms

Hazards
Disaster
Vulnerability
Hazards analysis
Vulnerability assessment
Consequence assessment
Risk

Issue

What processes will ensure that residents, government officials, and business leaders understand the nature of hazards and their impacts?

Introduction

Disasters are natural and human-caused events that have an adverse impact on a community, region, or nation. Events associated with a disaster can overwhelm response resources and have damaging economic, social, or environmental impacts. The capacity of the community, region, or nation to deal with the impacts of disasters provides a basis for us to classify the event as a crisis that can be addressed by local resources or a disaster that requires outside assistance and support. The process of clarifying the nature and impacts of hazards is the foundation of hazards analysis.

Over the past twenty-five years, we have seen escalating costs associated with the direct economic impacts of natural disasters. Although the number of injuries and causalities has been dropping in recent years, property damage has increased dramatically (Abramovitz 2001; Mileti 1999). Mileti notes that disaster losses have been increasing and will likely continue in the future (1999). He sees that damages will grow to an average of $50 billion annually—about $1 billion per week. Some experts believe that this is a relatively conservative estimate of losses, since there is little inclusion of indirect losses (i.e., loss of jobs, market share, productivity, etc.).

The rising cost of disasters has also paralleled the movements of our population to coastal regions, thus increasing their vulnerability to hazards. In addition, we have seen widespread adverse impacts of disasters in the form of massive displacement, economic losses, and suffering from all parts of our society. Hurricane Katrina clearly demonstrated that many members of our community suffered from the flooding and storm surge. Poststorm after action reports have consistently noted that governments at all levels were ill prepared to deal with such a massive disaster.

This book challenges us to first examine the nature of our community and the hazards that could impact our social, economic, and ecological systems. We go further to also identify an approach to develop a broad-based hazard risk management strategy to reduce risk and mitigate losses. This book provides a roadmap for clearly identifying hazards, our vulnerabilities, and both risk management and community-wide strategies for building sustainable communities.

Terminology of Hazards

Individuals interested in hazards and risks are exposed to multiple definitions of key terms such as "hazards," "disaster," "risk assessment," and "hazards analysis." Often, many of the terms are used interchangeably and, arguably, correctly or incorrectly. Many experts in the risk field have addressed the "terms" problem. For example, Kaplan describes two theorems of communication, which explain the confusion resulting from the different and often conflicting definitions of terms used in risk analysis and assessment (1997). The theorems state the following: Theorem (1), 50% of the problems in the world result from people using the same words with different meanings; Theorem (2), the other 50% comes from people using different words with the same meaning. This confusion has led organizations such as the Federal Emergency Management Agency (FEMA), the International Association of Emergency Managers (IAEM), the U.S. Environmental Protection Agency (EPA), and other federal agencies to increase the professionalism in the field by recognizing the need for a common set of definitions.

Hazards refers to a potential harm which threatens our social, economic, and natural capital on a community, region, or country scale. Hazards may refer to many types of natural (flood, hurricane, earthquake, wildfire, etc.), technological (hazardous materials spill, nuclear accident, power outage, etc.), or human-induced events (biochemical, bombing, weapons, mass destruction, terrorism, etc.). Compounded hazards are those that result from a combination of the above hazard types, such as urban fires resulting from earthquakes, failures of dams or levees resulting from flooding, or landslides resulting from wildfires and heavy rains.

FEMA describes hazards as "events or physical conditions that have the potential to cause fatalities, injuries, property damage, infrastructure damage, agricultural losses, damage to the environment, interruption of business, or other types of harm or loss" (1997). A hazard may be measured by its physical characteristics, likelihood, or consequences. For example, many coastal communities experience flooding. Water from heavy rains, levee breach, or dam break would be the source of the hazard. The likelihood could be considered a low risk or not likely; it could be a medium risk or one that has a high likelihood of occurring. A hazard has the potential to cause fatalities, injuries, property damage, infrastructure or agricultural loss, damage to the environment, interruption of business, or other types of harm.

Cutter notes that hazards evolve from interactions between natural, human, and technological systems (2001) but are also characterized by the areas of their origin. For example, the hazard may arise from a hurricane, but flooding may be magnified not only from excessive rainfall but also by long term nonsustainable agricultural or forest practices. Since a disaster could evolve from the interactions between social, natural, and technological systems, the classification of a complex hazard could be difficult. As a further illustration of the difficulty in classifying disasters, a hurricane or flood occurring in a community might also lead to an accidental release of a hazardous chemical from a container in flood waters. In this case, we have the potential of two disasters–one natural and the second human-caused or technological in nature. Cutter suggests that we view hazards within a broader social, political, historic, economic, and environmental context to fully appreciate how hazards can cause damage to community resources.

The United Nations defines a disaster as, "a serious disruption of the functioning of society, causing widespread human, material, or environmental losses which exceed the ability of affected society to cope using only its own resources" (United Nations 1992).

All disasters, small or large, are the result of a hazard being realized. There is a caveat to this definition, however, in that the realized hazard must overwhelm the response capability of a community to be considered disastrous (FEMA 1997).

Pearce (2000) suggests that any definition of disaster must reflect a given locality's capacity to respond. She goes on to state that the hazard event must be unusual and that the social, economic, political, and ecological impacts must be significant. She defines disasters:

> A disaster is a non-routine event that exceeds the capacity of the affected area to respond to it in such a way as to save lives; to preserve property; and to maintain the social, ecological, economic, and political stability of the affected region (p. 87).

Disasters are measured in terms of lives lost, injuries sustained, or property damaged and must be distinguished from routine emergency events that can result in property damage or fatalities. For instance, a house fire may require a response by a jurisdiction's fire department and result in loss of life or property. However, as fires are common emergency occurrences, they are managed by local response agencies and are normally not considered a disaster. For a fire to be considered a disaster, it must overwhelm the capacity of the local responders.

Common breakdown of hazards include atmospheric climatic hazards such as rain, lightning, wind and dust storms, hailstorms, snow avalanches, heat waves, snowstorms, and fog (Bryant 2005; FEMA 1997; Hewitt and Burton 1971). They also include geologic and seismic hazards such as landslides, avalanches, land subsidence, erosion, earthquakes, tsunamis, volcanic activity, and shifting sands. Hydrologic hazards make up the third type of natural hazard and include flooding, storm surges, coastal erosion, waves, sea ice, and sea level rise. Hewitt explains that compounded hazards include tropical cyclones, thunderstorms, whiteouts, tornadoes, rain and windstorms, blizzards, drought, freezing rains and wildfires; all combine several natural hazards and are not just the result of a single hazard.

Not all hazards result in disasters, for a hazard event could decrease potential damaging impacts so as to minimize losses (Gruntfest et al. 1978; Hewitt and Burton 1971; Lindell and Meier 1994). The rate or speed of onset of the event could give communities notice needed to minimize deaths and injuries by ordering an evacuation for a flooded area. Availability of perceptual cues (such as wind, rain, or ground movement) provide notice of a pending disaster. The intensity of a disaster could vary spatially so as to have damage impacts in areas with no social or economic impacts. Technology such as weather radar allows us to see where a heavy storm is moving so as to provide warning to the local area. The areal extent of the damage zone or its size (geographic area influenced) and its duration could influence any damaging impacts and the community's capacity to deal with the hazard event. Wind damage from a tornado could be limited to nonpopulated areas and not cause injuries or property damage. Finally, the predictability of the event or notice of occurrence is also critical in allowing those affected by the event to seek safety.

Despite our efforts to reduce our vulnerability to disasters, we see that property losses, deaths, and injuries continue to increase. Numerous studies have documented that increased losses are growing because (Abramovitz 2001; Mileti 1999):

1. Population growth occurs in high hazard areas.
2. Marginalized land is being developed, making us more susceptible to hazard impacts.
3. Larger concentrated populations in urbanized areas increase the potential for human and property loss; people are less familiar with hazards in their surroundings; growth may not be ecologically sustainable; more buildings and infrastructure may be damaged if an event occurs.
4. People are not impacted by hazards equally; economic disparities cause large numbers of impoverished people to be at risk.
5. Immense potential for loss as sea levels rise; weather and climate patterns will change.
6. Political unrest can directly cause loss (e.g., civil war) and/or make a region more susceptible to hazard impact due to lack of preparedness and/or inability to cope.
7. More property is at risk to hazards, but preparedness and mitigation measures minimize loss of human life.

The terms "risk" and "hazard"' are often used interchangeably and inconsistently. Different interpretations come from the fact that among emergency managers, risk managers, urban and regional planners, insurance specialists, and lay people, the meanings of the words have evolved independently and are used many different ways. These definitions can even be in conflict with each other. For example, it is not uncommon for the word risk to be used informally in a way that means "venture" or "opportunity," whereas in the field of risk management the connotation is always negative (Jardine and Hrudley 1997). However, even among risk managers, the exact definition of risk varies considerably (Kedar 1970).

The risk of disaster is typically described in terms of the probabilities of events occurring within a specified period of time, e.g., five, ten, or twenty years, a specific magnitude or intensity (or higher), or a range such as low, medium, or high risk. For example, the risk of floods is commonly described by FEMA in terms of 100- and 500-year floods, indicating the average frequency of major flooding over those periods of time and the maximum area that has been inundated each time. Risk has the common meaning of danger (involuntary exposure to harm), peril (voluntary exposure to harm), venture (a business enterprise), and opportunity (positive connotation—it is worth attempting something if there is potential for gain). In a business context, it refers to probability considerations, but is primarily concerned with uncertainty.

Views of Extreme Natural Events as Primary Causes of Disasters

Tobin and Montz (1997) provide a very insightful perspective on how we might view natural hazards and disasters. They see that one way of viewing disasters is that all or almost all responsibility for disasters and their impacts are attributed to the processes of the geophysical world. In this approach, the root cause of death and destruction is caused by extreme natural events rather than human interface with the environment. Under this view of disasters, those who suffer losses are seen as powerless victims who have limited control and simply react to the immediate physical forces and processes associated with disasters. The physical world is thus viewed as external force, separate from human actions. This perspective is noted by Burton and Kates (1964), who see natural hazards as those elements of the physical environment harmful to man and caused by forces external to him.

This perspective of disasters from Tobin and Montz is significant for the outputs of a hazards analysis. If the view of individuals in a high-risk area is just limited to the physical world, then there is little that can be done to minimize destructive hazard impacts. Quarantelli notes his early views of disasters and their origins.

> The earliest workers in the area, including myself, with little conscious thought and accepting common sense views, initially accepted as a prototype model the notion that disasters were an outside attack upon social systems that "broke down" in the face of such an assault from outside (1998: 266).

Steinberg (2000) also commented on this view of nature and disasters as extreme events which are beyond our control.

> ... [T]hese events are understood by scientists, the media, and technocrats as primarily accidents—unexpected, unpredictable happenings that are the price of doing business on this planet. Seen as freak events cut off from people's everyday interactions with the environment, they are positioned outside the moral compass of our culture (2000: xix).

This view of how people view hazards is clarified by the concept of "bounded rationality," inadequate information and ability to make sound choices in the face of risk. Tobin and Montz clarify its application to disasters by explaining that "bounded rationality" refers to the fact that "behavior is generally rational or logical but is limited by perception and prior knowledge" (1997: 5). Burton et al. note that:

> … [I]t is rare indeed that individuals have access to full information in appraising either natural events or alternative courses of action. Even if they were to have such information, they would have trouble processing it, and in many instances they would have goals quite different than maximizing the expected utility. The bounds on rational choice is dealing with natural hazards, as with all human decisions, are numerous (1978: 52).

A Changing Hazard Paradigm

Our efforts to understand the nature of hazards, their impacts, the likelihood of their occurrence, and how we use this information in hazard mitigation or other public policy decisions has resulted in alternative approaches to hazards analysis. FEMA and later the National Response Team (NRT) approach suggest a data-based quantitative emphasis on the characteristics of potential hazards, their likelihood of occurrence, and a prioritization of alternatives to address threats. A hazard in this context is viewed in single events with specific causal events.

A more quantitative approach to assessing risk was stressed by the National Research Council (1983). This approach has four elements, risk identification, dose response assessment, exposure assessment, and risk characterization. This model was used as the standard, beginning in 1980 as part of the Superfund legislation, and institutionalized as part of the evaluation of abandoned Superfund sites. This emphasis on quantitative analysis is also reflected in United Nations vulnerability and risk assessment processes (Coburn et al. 1994). Cutter (2003) notes that most risk assessments used probability estimators and other statistical techniques. EPA's approach was broadened in 1987 to look beyond just exposure as its carcinogenic potential to look at noncancer human health risk, ecological risk, and welfare risk. This process was characterized as a "relative-risk" approach that moved away from pollution control and technology fixes to one of risk reduction and sustainable approaches to pollution management (U.S. EPA 1987). A comparative risk analysis process is now used in EPA as a basis for environmental policy priority setting (Davies 1996) and an even broader examination of ecological risk is included in many risk assessments.

Cutter (2003) notes that these processes for risk assessment are fraught with methodology concerns that include uncertainty, especially with variability in individuals and ecosystems, and limited environmental data. Risk assessments must link good science with communities. This broader view of risk that includes communication and interaction between the scientist and those impacted by the assessment of hazards is very constructive.

Gaikie et al. examine disasters and their adverse impacts in two ways (1994). Two alternative models are provided to explain the complex nature of hazards and their impacts. A pressure and release model (PAR) examines the relationship between processes and dynamics that bring about unsafe conditions and their interface with disaster events such as earthquakes, floods, or tropical cyclones. The emphasis in the PAR are the driving social forces and conditions that bring about vulnerability of people in place. The second model emphasizes access to resources and takes into consideration the role of both political and economic conditions as the basic causes of unsafe conditions. Bull-Kamanga et al. stress the importance of local processes for risk identification and reduction (2003). A hazards analysis should include local players for it enhances the possibility of an accurate characterization of local conditions that may influence disaster impacts. This emphasis on underlying social conditions including hazards analysis is also stressed by Weichselgartner (2001). In his view, a disaster is a product of a cumulative set of human decisions over long periods, and these decisions either create greater risk or reduce risk. Mitigation must stress the underlying human conditions and not just adjust the physical environment. Vulnerability analysis thus must take a broad view of the conditions that are present prior to a disaster as well as the physical environment of a community and the characteristics of the hazard.

Critical Thinking: We build levees and flood walls, establish building codes and base flood elevations in an effort to protect property from hazards. How can hazards analysis help to foster greater awareness of risks associated with hazards? How can a hazards analysis have a constructive influence on decision-making at the individual, family, and community levels?

Cutter (1996, 2003) also suggests a broader perspective is needed to fully appreciate the complexities associated with hazards and their numerous impacts. This hazards of place model of vulnerability involves a comprehensive understanding of hazard potential along with an examination of the geographic context, social conditions, and both biophysical and social vulnerability. A place vulnerability is then determined from these elements. A set of indicators can be used to examine vulnerability and take into account population variables and infrastructure lifelines. The goal in this approach is to assess social vulnerability and community sustainability for response and recovery. Cutter stresses that, in order to understand a community's hazard potential, one must consider that it is influenced by socioeconomic indicators, individual characteristics, and the community's geographic context impacted by the hazard (1996). Others have also developed criteria to examine community sustainability. Miles and Chang (2006) examine community recovery capacity by using social and infrastructure variables. They stress the need for modeling recovery processes in understanding community resiliency capacity.

An emphasis of modeling hazards and measuring the resilience of communities is also seen in Bruneau et al. (2000). The quantitative measures combined with characterization of a hazard results in information that may be used in guiding mitigation and preparedness efforts. Their measurement of the local community centers around four characteristics, including robustness of systems to withstand loss of function, redundancy of elements of the system to suffer loss, resourcefulness and the capacity to mobilize resources when the system is threatened, and rapidity or the speed to achieve goals in a timely manner.

Hazards Analysis

There are many perspectives on "hazard analysis" that vary from FEMA's approach of knowing what could happen, the likelihood of it and having some idea of the magnitude of the problems that could arise (FEMA 1983). FEMA's introduction of the HAZUS modeling software in 1997 reflects their interest in physical processes at the community or regional level (1997). This approach unfortunately is limited to modeling one hazard at a time and fails to address a multihazard environment or multiple risks (Cutter 1996). A process approach to hazards analysis addresses adverse impacts of hazards (Long and John 1993) and stresses the role of hazard identification, risk screening, and the development of mitigation measures to control losses. The Coastal Engineering Research Center and the University of Virginia, along with the United States Geological Survey (USGS), used a quantitative hazards analysis approach in examining risk and exposure to coastal hazards (Anders et al. 1989). They evaluated the U.S. coastline for risk and exposure to coastal hazards by examining the characteristics of the hazards, coastal geographic features, population demographics, and civic infrastructure.

Researchers from Oak Ridge National Laboratory identified a coastal vulnerability index that includes risks from sea level rise in coastal communities (Daniels et al. 1992; Gornitz and White 1992; Gornitz et al. 1994). This study weighed the characteristics of coastal hazards, local geographic conditions, and the likelihood of extreme weather impacting local areas. Their approach to hazards analysis focused on characteristics of the hazard event, local geographic conditions, and demographic factors. They emphasized the use of model outputs in mitigating disaster impacts. Multihazard impacts have been examined by Preuss and Hebenstreit (1991) to understand the impacts of both earthquakes and associated tsunami flood events. This risk-based urban planning approach was designed to allow assignment of risk factors for vulnerabilities on a community basis.

The Environmental Protection Agency along with fourteen other Federal Agencies (National Response Team, NRT) adopted a common approach to community level hazards analysis and planning (NRT 1987). This approach uses a

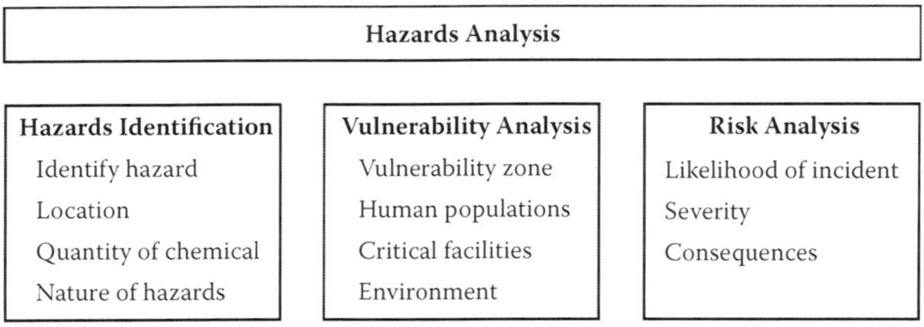

Figure 1.1 U.S. Environmental Protection Agency hazards analysis process.

process format in providing communities with a broad understanding of hazards and risks. NRT-1 defines hazard analysis as a three-step process: (1) hazard identification, (2) vulnerability analysis, and (3) risk analysis. This approach to hazards analysis stresses the need for broad-based information to support community decision making to reduce vulnerability and minimize risk to people and property (Figure 1.1).

Critical Thinking: How do you ensure that an analysis of hazards and their potential adverse impacts is sensitive to the human, cultural, economic, political, and natural environments and not limited to the built environment?

Hazard Identification

Hazards identification provides specific information on the nature and characteristics of the hazardous event and the community. It further examines an event's potential for causing injury to life or damage to property and the environment. Hazard identification takes advantage of the use of environmental modeling to characterize hazards and disaster impacts. As part of the EPA hazards analysis process, community involvement is encouraged through a broad-based team represented by local response agencies, the media, community public health units, medical treatment organizations, schools, public safety, and businesses. The formation of local emergency planning committees provides the basis for broad input in preparedness efforts.

Vulnerability Analysis

Vulnerability analysis is a measure of a community's propensity to incur loss. Vulnerability analysis may focus on physical, political, economic, and social vulnerability. Vulnerability is, in other words, the susceptibility to hazard risks. Vulnerability can also be a measure of resilience. According to Emergency Management Australia (2000), vulnerability is "The degree of susceptibility and resilience of the community and environment to hazards." Vulnerability analysis identifies the geographic areas that may be affected, individuals who may be subject to injury or death, and what facilities, property, or environment may be susceptible to damage from the event:

1. The extent of the vulnerable zones (i.e., an estimation of the area that may be affected in a significant way).
2. The population, in terms of numbers, density, and types of individuals (e.g., employees; neighborhood residents; people in hospitals, schools, nursing homes, prisons, and day care centers) that could be within a vulnerable zone.

Vulnerability analysis examines who and what is vulnerable and why:

What private and public property (e.g., critical facilities, homes, schools, hospitals, businesses, offices) may be damaged, including essential support systems (e.g., water, food, power, communication, medical) and transportation facilities and corridors? What property and groups are more likely to be affected by a disaster?

How could the environment be affected? What impact could a disaster have on sensitive natural areas and endangered species?

What social groups are likely not to receive or not to understand or not to take the warning message seriously? Why?

Are there characteristics of social groups that may make it more difficult for them to be rescued, to receive adequate emergency medical care, to feel comfortable in an emergency shelter? Are there population groups that are likely to suffer to a great extent economically or emotionally in a response as well as in a recovery?

When assessing risks, experts must factor in vulnerability. The vulnerability assessment is a measure of the exposure or susceptibility and resilience of a community to hazards. We stress that understanding vulnerability by itself is insufficient to plan for disasters. It must be accompanied by understanding the nature and characteristics of hazards. Hazards identification and characterization is thus a component of a full hazards analysis.

Risk Analysis

Risk analysis is described by the EPA as an assessment of the likelihood (probability) of an accidental release of a hazardous material and the consequences that might occur, based on the estimated vulnerable zones. The risk analysis is a judgment of probability and severity of consequences based on the history of previous incidents, local experience, and the best available current technological information. It provides an estimation of:

1. The likelihood (probability) of a disaster based on the history of current conditions and consideration of any unusual environmental conditions (e.g., areas in flood plains), or the possibility of multiple incidents such as a hurricane with tornadoes (e.g., flooding or fire hazards).
2. Severity of consequences of human injury that may occur (acute, delayed, and/or chronic health effects), the number of possible injuries and deaths, and the associated high-risk groups.
3. Severity of consequences on critical facilities (e.g., hospitals, fire stations, police departments, communication centers).
4. Severity of consequences of damage to property (temporary, repairable, permanent).
5. Severity of consequences of damage to the environment (recoverable, permanent).

Risk in this view is the product of the likelihood of a hazard occurring and the adverse consequences from the event. Simply stated,

$$\text{RISK} = \text{LIKELIHOOD} \times \text{CONSEQUENCE}$$

Reality check: Increasing numbers of people are moving into vulnerable areas. This creates stress on resources and land use. Larger numbers of people may move into more sensitive environments. The best land may already be developed and in use, leaving available for development only areas that are marginal and more susceptible to hazards.

Linking Hazards Analysis to Risk and Comprehensive Emergency Management

Alexander describes two approaches to dealing with risks: one is a community hazard mitigation approach based on a comprehensive hazards analysis, large-scale planning, and decisions at the local community level. A second approach involves extensive risk communication with the community. In this approach, he suggests that we establish a greater understanding of local hazards and risk by the public. He suggests that communities include citizens in the hazards analysis process and proceed from nonstructural to structural protection, not vice versa (2000: 27).

Alexander suggests that decision-making at the individual, household, neighborhood, organizational, or community level can be made by informed individuals; rational choices may be based on information from a comprehensive hazards analysis. Risk communication is thus a critical part of the hazards analysis process and a positive contribution to decision-making processes. Through risk communication and public participation in the hazards analysis process, risk management and hazard mitigation strategies may be adopted by both citizens and the community to reduce vulnerability.

Individual citizens and communities thus make decisions that either increase or reduce our vulnerability to hazards. The key is to acknowledge the interface between environmental hazards and human actions and that actions can be initiated to reduce vulnerability through hazard mitigation and hazards risk management. As a result, the hazards analysis process must include opportunities for public involvement and risk communication and recognize that decision making is essential in adopting effective risk management and hazards mitigation strategies by individuals, organizations, and a community. Through this approach we reject a perspective that adopts the causality of environmental determinism where we have no power to reduce our vulnerability to environmental hazards.

This comprehensive approach to hazards analysis views disasters and hazards beyond just their geophysical processes and examines how social, economic, and political processes impact hazardousness. The importance of a community assessment as a part of hazards identification, which includes a close look at social-cultural, economic, and political systems, may be seen vividly in the impacts from Hurricane Katrina in New Orleans. The damages from Hurricane Katrina revealed significant vulnerabilities associated with poverty, education, housing, employment, and governance. Unfortunately, what was revealed in New Orleans is present in many urban coastal cities. In fact, Mileti stresses the need for a community assessment and a clear delineation of hazard areas within the hazards analysis process.

> (R)ebuilding that generally keeps people and property out of harm's way is increasingly viewed as an essential element of any disaster recovery program. Rebuilding that fails to acknowledge the location of high-hazard areas is not sustainable, nor is housing that is not built to withstand predictable physical forces. Indeed, disasters should be viewed as providing unique opportunities for change–not only to building local capability for recovery–but for long-term sustainable development as well.
>
> **Mileti (1999: 237–238)**

Figure 1.2 shows how the hazards analysis process builds on the U.S. EPA approach of hazard identification, vulnerability assessment, and risk analysis by stressing the need for the use of the results of a risk analysis in hazard adaptation adjustments. Risk communication, citizen participation, problem solving, risk management, hazard mitigation, and ongoing assessment are all parts of comprehensive emergency management. This approach stresses an action orientation through the adoption and implementation of comprehensive hazard risk management and hazard mitigation strategies and monitoring the effectiveness of hazard adjustments that are adopted and implemented. The ultimate goal of these hazard adjustments is to build resilient and sustainable organizations and communities.

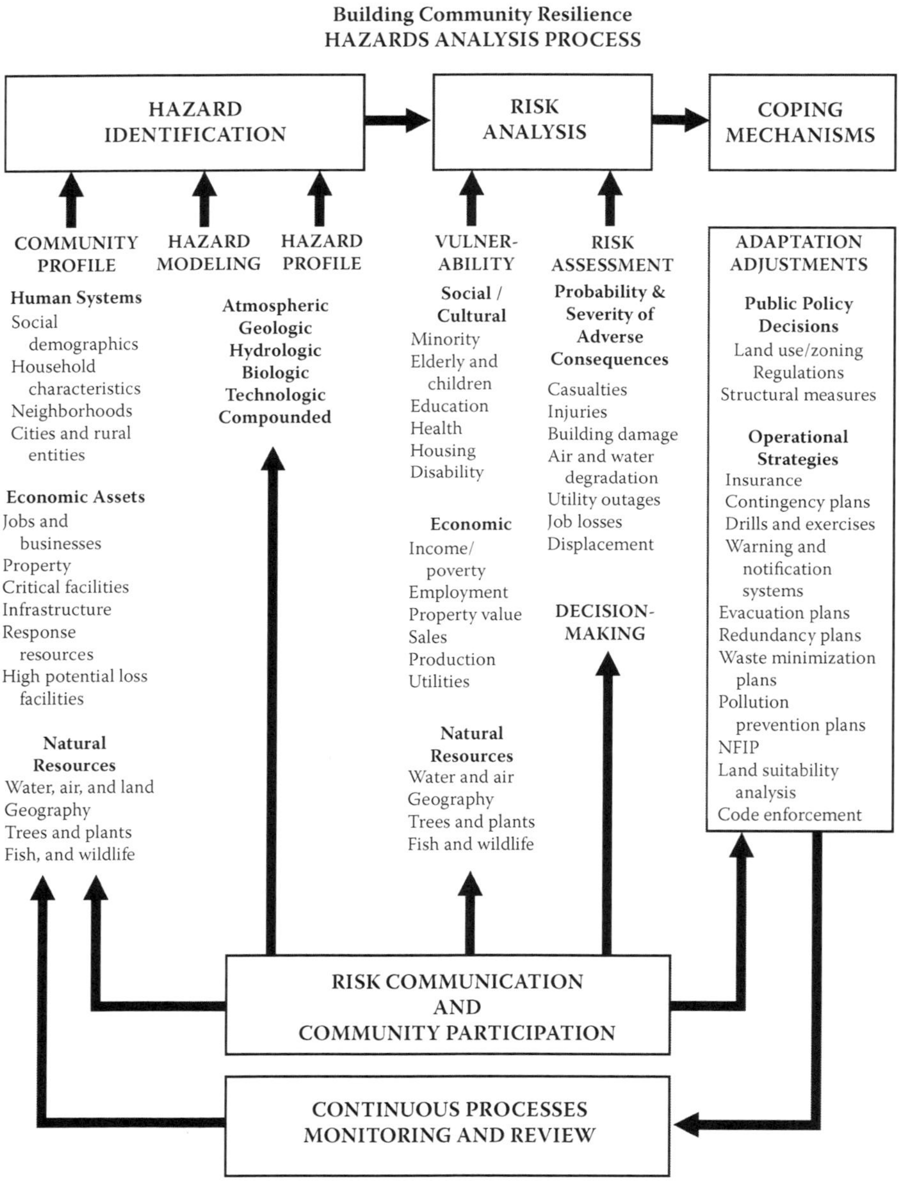

Figure 1.2 Hazards analysis process in building community and organizational resilience. Graphic design by Mary Lee Eggart.

Through hazard adaptation and adjustments, we stress the role of risk communication and community and organizational participation in the hazards analysis process. Understanding of hazards will not be the result of just telling people of hazards, but also of allowing them to participate in the hazards analysis process at the neighborhood, community, and organizational level. People support what they help build, and citizens as well as employees will advocate risk management and hazard mitigation strategies that they understand and help formulate. They will likely oppose what is imposed on them.

Further, we stress that the hazards analysis process includes an intentional assessment or monitoring of the impact of our hazard mitigation strategies and hazard risk management strategies. We want to know what are the short- and long-term results of our actions, which might include increasing minimum base flood elevation requirements, strengthening building code requirements, or enhancing building inspections. Second, we stress that the hazards analysis process is not static but an ongoing one. We see that the ongoing review of our hazard mitigation and risk management policies could lead to program changes to strengthen or enhance opportunities to build more sustainable and hazard resilient communities and organizations.

Critical Thinking: One of the best hazards analysis efforts that has been conducted on a large scale was completed in 2007 by the Coastal Protection and Restoration Authority (CPRA 2007) of Louisiana (http://www.lacpra.org/).

The CPRA plan shown in Figure 1.3 included a comprehensive assessment of the hazards in a coastal environment, impact assessments (social, economic, and ecological), public participating, and recommendations on strategies to protect the social, economic, and environmental assets of the state. This is an outstanding example of a region-wide hazards analysis which included federal, state, and local government agency collaboration.

Communicating Risk from a Hazards Analysis

The results of a hazards analysis are not just for planning and mitigation of hazards but should be shared with the public. All communities have risks associated with natural hazards. Hazards that typically cause minimal damage are usually accepted as inevitable, and little is done to reduce the risk. Such hazards may be viewed as nuisances, rather than real threats to life and property (Waugh). Some communities are willing to accept more risk than others. Factors such as the political culture and the socioeconomic level of the community determine the levels and kinds of risk that may be accepted. For example, poor communities may be willing to accept more risk from environmental hazards, because the economic base of the community will not directly support the allocation of resources for structural or nonstructural hazard mitigation initiatives. Individual residents may refuse to purchase flood insurance or take other measures that have associated costs, because

Figure 1.3 (See color insert following page 142.) Louisiana's Comprehensive Master Plan for a Sustainable Coast (http://www.lacpra.org/).

they have limited discretional financial resources. They want the insurance but just do not have the funds.

Communicating information from the hazards analysis to the public can help shape perceptions of risk and elevate concern for protecting personal property. Further, by acknowledging local environmental risks, the community may initiate strategies that can overcome the individual financial limitations so as to protect the entire community from hazards. Collective action may be advisable when low-income residents who may be renters or homeowners cannot take individual action. Individual risk assessment, risk management, and impact assessment are all part of using information from a hazards analysis to protect individual citizens and their property.

Community Involvement in Hazards Analysis

The approach used in a hazards analysis can include a broad-based representative methodology that ensures that outputs address community priorities and concerns. Some agencies and scientists see hazards analysis as a scientific process that includes only the experts. We suggest an alternative approach. Partnering experts in community planning, engineering, modeling, geography, sociology, or other hazard fields with community leaders and members will establish a dialogue relating to risks to reach common goals around generating insights and strategies to build a sustainable community (Stringer 1999). It operates through a dialogic, hermeneutic approach, similar to fourth-generation evaluation (Lincoln and Guba 1989). An external group of experts can be well positioned to collect the right information and save local governments time in dealing with unknown areas. Engagement seeks a grass-roots understanding of risk, one that is perceived as critical to risk reduction and building local capacity (Heinz Center 2002). This report suggests that the engagement process should include the following:

1. Experts (public agency representatives, including emergency management and other local agencies along with consultants, if used) meet with local residents to explore common goals in a hazards analysis.
2. Identify questions and issues relevant to the residents, including the roles of residents or community members and agency experts (outsiders).
3. Develop, through consensus-building, common objectives and priorities for the hazards analysis beneficial to both the experts and residents.
4. Describe the hazards analysis process.
5. Develop, through consensus-building, an agreed-upon strategy of how the results of the hazards analysis will be shared with the community, organizations, and public officials.
6. Discuss residents' concerns.
7. Initiate the project(s).
8. Present the results to the community for their response.

Values in Community Engagement

Greenwood and Levin (1998) suggest that this approach is context bound and deals with real-life problems. It is problem focused and joins participants and experts to generate through collaborative discussion. All participants have meaningful dialogue, diverse points of view are welcomed, and the process leads to action. The key is that actions evolve to address problems associated with hazards, and both the experts and citizens have an increased awareness of options to address problems.

One of the main values in community engagement is community sustainability, characterized by environmental quality, quality of life, disaster resiliency, vibrant economies, and equity as developed through local consensus building (Mileti 1999). Sustainability implies persistence within an ecosystem's carrying capacity (Burby 1998). Systems theory guides the analysis by examining the natural, human, economic, political, and constructed systems. "Hazards researchers and practitioners would do well to take a more systems-based approach. … [it] recognizes multiple and interrelated causal factors, emphasizes process, and is particularly interested in the transitional points at which a system … is open to potential change" (Mileti 1999: 106–107). From this standpoint, hazards are viewed as "complex interactions between natural, social, and technological systems" (Cutter 1994: xiv). Those interactions result in vulnerability (Heinz Center 2000). To assess those systems and their interactions, the analysis should include an examination of the following.

Human Social and Cultural Heritage Elements

This part of the analysis attends to culture, ethnic identity, social institutions (family, faith, economy, education, self-governance), and disaster experiences. Members of the community are encouraged to share family and community photographs and stories related to the human, natural, and physical systems.

Protective Actions

Risk communication (warnings), land use and zoning (life and property), and participatory community meetings will involve residents in hazard identification, risk and vulnerability assessment, and planning. Data collection will center on disaster history, vulnerability, and sociobehavioral response. This focus will address interactions between the natural and human systems. We are especially interested in protective actions taken during several phases of disaster, with a particular concentration on social bonds.

Constructed and Physical Environment

This part of the analysis examines how the constructed environment links to the human and natural environment. This focus examines how the community's sense of place mediates their relationship to the social, physical, and built environment. Data from a risk assessment is included, such as hurricane wind and storm modeling, riverine flood hazard modeling, and earthquake, wildfire, and wind models to examine potential social, economic, and environmental vulnerabilities. This focus addresses the built and natural systems as experienced within the human system.

Critical Thinking: Is the concept of community right to know just a legal obligation, or is it based on a broader set of values? The adoption of the Community Planning Right to Know Act (U.S. EPA 1986) changed how local communities conducted hazards analysis and communicated information about hazards to the community. This legislation asserts that the community has a right to know of chemical hazards present in the community. Although this legislation focused on human-caused technological hazards, it was built upon shared societal values. The "right to know" implies other values: transparency, accountability, responsible action, democracy, and active citizen participation. Right to know places on emphasis on risk communication and sharing of information freely about hazards. This however changed following September 11, 2001. Since that time, new restrictions have been established relating to access to chemical hazard information "in the interest of security."

Conclusion

In this book, we wish to suggest that this broader approach to understanding hazards must be used. Natural hazards have very different impacts throughout the world, and the nature and extent of a disaster depends of several factors including (1) local and regional environments including the landscape, climate conditions including the probability of an event, how often they occur, and the capacity of the hazard to do harm; (2) the strength and vitality of the social, economic, and natural environments to withstand and cope with the adverse effects of a hazard; and (3) response and recovery resources that enable communities, regions, and nations to cope and recover from disasters. The capacity of the organization, community, and region to recognize their vulnerability and initiate steps to reduce adverse impacts is critical to the hazards analysis process.

In addition to viewing hazards and disasters as part of a local condition, we wish to stress the use of hazards analysis for mitigation and prevention rather than just for response and recovery or as part of the regulatory permitting process. The results of a hazards analysis can provide information to identify hazard risk management strategies to strengthen social, economic, and environmental systems and enable these systems to withstand the destructive conditions that are inherent in hazards. It may be impossible to reduce the wind or storm surge from a hurricane or the shaking from an earthquake, but we can take steps to build stronger buildings, locate our structures in less vulnerable areas, and enhance our social structures to cope with displacement, loss of jobs, and critical natural resources.

We link the hazards analysis process to decision-making by local community officials, individual citizens, and private and nonprofit organizations. This emphasis on decision making is reflected in Deyle et al. (1998), where hazard assessments provide a factual rational basis for local decision making. Their goal is to achieve safer more sustainable communities through management and informed decisions based on estimates of costs and benefits of efforts to reduce risks. The approach suggested in this text builds on the EPA approach of hazard identification, vulnerability assessment, and risk analysis, but emphasizes decision making within social-cultural, political, and legal constraints. We stress that the scale of analysis is critical in establishing the type of data that is needed and the degree of precision that may be needed in a community, and that data used must be current.

Lindell and Prater (2003) explain that the hazards analysis process is linked to comprehensive emergency management and hazard risk management processes through a disaster impact model. Their view suggests that there are existing conditions that reflect current hazards as well as current social and economic vulnerabilities. In addition, they note that the current physical conditions of the community will impact its resilience. They contend that the social and economic impacts from a disaster are greatly influenced by the community's level of implementing effective mitigation strategies, emergency preparedness, and recovery strategies as well as the physical characteristics of the disaster and the community's actual response and recovery efforts. This broad view of the hazards analysis process acknowledges the need to understand the nature of the community and develop a broad community profile that includes an examination of local geography, demographics, infrastructure, and response resources. They go further to stress the importance of using hazards analysis in the preparation and monitoring of hazard mitigation strategies, emergency preparedness plans, and response strategies.

The nature and extent of a hazard condition thus influences the potential adverse impacts for the built and human environments. Wildfires that are driven by 30-mph winds have a greater destructive force than fires in low wind conditions. Each of these hazardous conditions has very different impacts on any community depending on local conditions including:

The character of our built environment such as homes, office buildings, manufacturing plants, roads, bridges, dams, or levees

The nature and condition of the natural environment such as wetlands, flood plains, forests, cultivated areas, hills, mountains, or changes in elevation

The existence of strong and connected families, neighborhood associations, nonprofit groups, and individuals who are engaged in their community

Communities with high unemployment, poverty, excessive crime, and poor education may have great difficulty in coping and recovering from a hazardous event. Communities vary in the resources that may be used to deal with a hazardous event. Some communities may have large numbers of the local population living in poverty and high unemployment, high crime, and limited pubic resources to deal with a disaster event. The design of built structures and the nature of natural environments can influence the damage that results from a hazard event. Healthy wetlands and forests, strengthened structures, and land that is on higher ground are in a better position to withstand the destructive character of hurricane winds or flooding conditions. Existing conditions may be assessed as the community's capacity to resist damage from disasters.

Measuring a community's resilience has been examined by a growing number of hazard researchers (Bruneau et al. 2000; Tierney and Bruneau 2007). One must look at numerous local systems that are exposed to hazards (Turner et al. 2003). Vulnerability should be considered not just by exposure to stresses produced by hazards alone, but also by the sensitivity and resilience of the social-cultural, economic, and environmental systems impacted by the hazards. They stress that vulnerability assessment must examine coupled human and environmental systems and their linkages within and without the systems that affect their vulnerability. Turner et al. present a framework for the assessment of coupled human and environmental systems (2003).

A comprehensive hazards analysis must take this broad approach and acknowledge the human, built, and natural environmental systems and their multiple connections. Hazards analysis must attempt to understand each of these systems and examine their interconnectedness and impact on community resilience.

We do not suggest a new approach or methodology in hazards analysis in this book, but we do stress the need to embrace a broader context in which to view the hazards analysis process. We agree with Cutter's observation that there is a need for a more integrative approach in vulnerability science for understanding and responding to environmental hazards (2003). Mileti (1999) notes that a new paradigm is needed in dealing with hazards and disasters, one that addresses sustainable hazard reduction. To accomplish this, the following should be addressed:

- *Sustainable culture*: We do not control nature despite our efforts to design levees, dams, or buildings, and in many cases we are the cause of disaster losses. We must understand the nature of hazards and build to reduce losses. The outputs of the hazards analysis process must be used to identify mitigation strategies so as to minimize our vulnerabilities socially, economically, and environmentally.
- *Events, losses, and costs*: Outputs from the hazards analysis process need to characterize our vulnerability and document how disasters have affected our communities.
- *The interactive structure of risk*: The hazards analysis process can characterize our vulnerability and quantify areas that could be affected by a disaster. This process must also provide a broader view so we can see the social, political,

economic, and environmental costs to our communities. This broader view of risk allows us to include many different interest groups in making decisions about reducing our vulnerability.

- *Land-use management*: Local decision makers can use outputs of a hazards analysis in land-use plans. Limiting development may contribute to the social, ecological, and economic sustainability of our communities.
- *Engineering codes and standards*: Local government adoption of codes and enforcement process are critical in reducing our vulnerability to disasters. A comprehensive hazards analysis provides critical information to ensure that code enforcement goals are attained.
- *Prediction, forecast, warning, and planning*: A detailed hazards analysis provides a sound basis for ensuring that local communities can offer citizens adequate disaster warning. Procedures for delivering timely warning for disasters can be based on alternative planning scenarios from a hazards analysis.
- *Disaster response and preparedness*: Emergency preparedness plans are prepared on a comprehensive hazards analysis. Policies and operational procedures are driven by the nature of the hazards faced by organizations and communities.
- *Recovery and reconstruction*: Planning for recovery should not begin following a disaster. To be effective, it should be part of a community hazard mitigation plan and include priorities for a community's long-term recovery in the event of a disaster.
- *Insurance*: Insurance is not a prevention strategy, but it can be included as part of a recovery process. The question is, how can we use insurance as a means of ensuring that an entity's financial stability is protected?
- *Economic sustainability*: Public, private, and nonprofit organizations must understand the nature of risks facing them and develop strategies to reduce or eliminate losses. A hazards analysis is critical to this decision-making process.

Deyle et al. (1998) stress the application of a community hazards analysis in local or regional decision making and land-use planning. Deyle et al. agree with the suggestions of Meleti above that analysis without action does not address the critical decision that must include a comprehensive understanding of hazards, vulnerability, and risk. Hazards analysis is part of a comprehensive emergency management and risk management process. Hazards analysis is not the goal but is a means toward a goal of promoting social, economic, and environmental sustainability. This emphasis is on hazard risk management and mitigation so as to foster community and environmental sustainability and resiliency.

Discussion Questions

The science in hazards analysis. (Why do we worry when the scientist says that there is no risk?)

Why are hazard analyses necessary? (In what way does hazard analyses provide information on the known environmental risks in the community or region? Does the community have a right to know about local hazards?)

What is the role of hazards analysis in organizational decision making and public policy?

How are disasters and development related? What is the role of a hazards analysis in preventing people from moving into harm's way?

What are the implications of this in terms of vulnerability and exposure to hazards?

Much of the hazards research documents increasing damage from disasters. How do demographic, economic, political, and environmental change contribute to vulnerability? How could a better understanding of hazards and the impacts from disasters help us to reduce the adverse consequences associated with disasters?

What type of demographic, economic, political, and environmental changes could make a community more resilient and less vulnerable to disasters?

Do you agree that citizens have a right to know about hazards in their community? Under what conditions might a community restrict information about hazards?

How are disasters and development related? How is resilience related to hazards analysis?

Applications

Take a look at the hazards analysis process outlined in Figure 1.2. Identify examples of how population characteristics, the local economy, the infrastructure in the community, and the natural environment influence the community's vulnerability to natural hazards. What hazards appear, at this first look, to be the primary threat for your community and should be addressed through a comprehensive hazards risk management and mitigation strategies?

Web Sites

Coastal Protection and Restoration Authority (CPRA) of Louisiana (2007). *Integrated Ecosystem Restoration and Hurricane Protection: Comprehensive Master Plan for a Sustainable Coast.* http://www.lacpra.org/index.cfm?md=pagebuilder&tmp=home&nid=24&pnid=0&pid=28&fmid=0&catid=0&elid=0. The plan includes a comprehensive assessment of the hazards in a coastal environment, public participating, and recommendations on strategies to protect the social, economic, and environmental assets of the state. This is an outstanding example of a region-wide hazards analysis, which included federal, state, and local government agency collaboration.

FEMA (1997). *Multi-Hazard Identification and Risk Assessment: The Cornerstone of the National Mitigation Strategy.* Federal Emergency Management Agency, Washington, DC. http://www.fema.gov/library/viewRecord.do?id=2214.

NOAA (2007). Coastal Services Center. Vulnerability Assessment Techniques and Applications (VATA). National Oceanic and Atmospheric Administration, Coastal Services Center. http://www.csc.noaa.gov/vata/links.html.

NOAA (2007). Risk and Vulnerability Assessment Tool (RVAT). National Oceanic and Atmospheric Administration, Coastal Services Center. Hazards Analysis, an easy-to-use, adaptable, multi-step process that includes the hazard identification. http://www.csc.noaa.gov/rvat.

NRT (2001). Hazardous Materials Emergency Planning Guide. NRT-1 2001 Update. U.S. Environmental Protection Agency, National Response Team. Planning, preparedness, and response actions related to oil discharges and hazardous substance releases. http://www.epa.gov/region6/6sf/pdffiles/nrt_1_planning_guide.pdf (PDF).

U.S. Army Corps of Engineers (2007). New Orleans Risk and Reliability Report. U.S. Army Corps of Engineers. http://nolarisk.usace.army.mil.

U.S. EPA (1987). *Technical Guidance for Hazards Analysis: Emergency Planning for Extremely Hazardous Substances.* U.S. Environmental Protection Agency. http://www.epa.gov/emergencies/docs/chem/tech.pdf (PDF).

USGS (2008). United States National Seismic Hazard Maps. United States Geological Survey. Learn more about seismic hazard analysis, http://earthquake.usgs.gov/hazmaps/.

References

Abramovitz, J. (2001). *Unnatural Disasters.* Worldwatch Paper 158. World Watch Institute. Washington, DC.

Alexander, D. (2000). *Confronting Catastrophe.* Oxford University Press, New York, 23–36.

Anders, F., S. Kimball, and R. Dolan (1989). *Coastal Hazards: National Atlas of the United States.* U.S. Geological Survey, Reston, VA.

Blaikie, P., T. Cannon, I. Davis, and B.Wisner (1994). *At Risk.* Routledge, New York.

Bruneau, M., S. E. Chang, R. T. Eguchi, G. C. Lee, T. D. O'Rourke, A. M. Reinhorn, M. Shinozuka, K. Tierney, W. A. Wallace, and D. von Winterfeldt (2000). A framework to quantitatively assess and enhance the seismic resilience of communities. *Earthquake Spectra* 19(4):733–752.

Bryant, E. A. (2005). *Natural Hazards.* Cambridge University Press. Cambridge, UK.

Bull-Kamanga, L., K. Diagne, A. Lavell, E. Leon, F. Lerise, H. MacGregor, A. Maskrey, M. Meshack, M. Pelling, H. Reid, D. Satterthwaithe, J. Songsore, K. Westgate, and A. Titambe (2003). From everyday hazards to disasters: the accumulation of risk in urban areas. *Environment and Urbanization* 15(1):193–204.

Burby, R. (1998). Cooperating with nature: confronting natural hazards with land-use planning for sustainable communities. Joseph Henry/National Academy Press: Washington, DC.

Burton, I. and R. W. Kates (1964). The perception of natural hazards in resource management. *Natural Resources Journal* 3; 412–441.

Burton, I., R. Kates, and G. White (1978). *The Environment as Hazard.* Oxford University Press, New York, 81–90.

CPRA (2007). Integrated Ecosystem Restoration and Hurricane Protection: Louisiana's Comprehensive Master Plan for a Sustainable Coast. Coastal Protection and Restoration Authority of Louisiana, Baton Rouge, LA.

Coburn, A. W., R. J. S. Spence, and A. Pomonis (1994). *Vulnerability and Risk Assessment.* United Nations Development Program. United Nations. New York.

Cutter, S. L. (1993). *Living with Risk.* Edward Arnold, London.

Cutter, S. L. (ed.) (2001). *American Hazardscapes: The Regionalization of Hazards and Disasters*. Joseph Henry Press, Washington, DC.

Cutter, S. L. (2003). The vulnerability of science and the science of vulnerability. *Annals of the Association of American Geographers* 93(1):1–12.

Davies, J. (1996). Comparing environmental risks: tools for setting government priorities. Resources for the future. Washington, DC.

Deyle, R. E., S. French, R. Olshansky, and R. Paterson (1998). Hazard assessment: the factual basis for planning and mitigation. In *Cooperating with Nature: Confronting Natural Hazards with Land-Use Planning for Sustainable Communities*, ed. R. Burby. Joseph Henry Press. Washington, DC.

Emergency Management Australia (2000). *Emergency Risk Management: Applications Guide.* Australian Emergency Manual Series. Emergency Management Australia, pp. 10–12. http://www.ema.gov.au/agd/EMA/rwpattach.nsf/VAP/(383B7EDC29CDE21FBA276BBBCE12CDC0)~Manual+05a.pdf/$file/Manual+05a.pdf.

FEMA (1997). *FEMA's Multi-Hazard Identification and Risk Assessment: The Cornerstone of the National Mitigation Strategy.* Federal Emergency Management Agency, Washington, DC. http://www.fema.gov/library/viewRecord.do?id=2214.

Gornitz, V. M. and T. W. White (1992). A Coastal Hazards Data Base for the U.S. East Coast. Oak Ridge National Laboratory, Environmental Sciences Division, Publication No. 3913 and 4101. Oak Ridge, TN.

Gornitz, V. M., T. W. White, R. C. Daniels, and K. R. Birdwell (1994). The development of a coastal risk assessment database: vulnerability to sea level rise in the U.S. Southeast. *Journal of Coastal Research,* Special Issue No. 12. Coastal Hazards: Perception, Susceptibility and Mitigation, 327–338.

Greenwood, D. and M. Levin (1998). *Introduction to Action Research: Social Research for Social Change.* Sage, Thousand Oaks, CA.

Gruntfest, E., Downing, T., and White, G. F. (1978). Big Thompson flood exposes need for better flood reaction system. *Civil Engineering 78*:72–73.

Heinz Center (2000). The Hidden Costs of Coastal Hazards. Island Press: Washington DC.

Hewitt, K. (1998). Excluded perspectives in the social construction of disaster. In *What is a Disaster? Perspectives on the Question*, ed. E. L. Quarantelli. Routledge, New York.

Hewitt, K. (1997). *Technological Hazards. Regions of Risk: A Geographical Introduction to Disasters.* Longman, Essex, England.

Hewitt, K. (1983). The idea of calamity in a technocratic age. In *Interpretations of Calamity from the Viewpoint of Human Ecology*, ed. K. Hewitt. Allen and Unwin, Boston.

Hewitt, K. and I. Burton (1971). The Hazardness of a Place: A regional ecology of damaging events. University of Toronto Press, Toronto.

Jardine, C. G. and S. Hrudley (1997). Mixed messages in risk communications. *Risk Analysis* 17(4):489–498.

Kaplan, S. (1997). The words of risk analysis. *Risk Analysis* 17(4):407–417.

Kedar, B. Z. (1970). *Again: Arabic Risq, Medieval Latin Risicum, Studi Medievali.* Centro Italiano Di Studi Sull Alto Medioevo. Spoleto.

Lincoln, Y. and E. Guba (1989). Fourth Generation Evaluation. Sage Publications: Thousand Oaks, CA.

Lindell, M. K. and M. J. Meier (1994). Planning effectiveness: effectiveness of community planning for toxic chemical emergencies. *Journal of the American Planning Association* 60(2):222–234.

Lindell, M. K. and C. S. Prater. (2003). Assessing Community Impacts of Natural Disasters. *Natural Hazards Review* 4(4) 176–185.

Liverman, D. M. (1990). Vulnerability to global environmental change. In *Understanding Global Environmental Change: The Contributions of Risk Analysis and Management*, eds. R. E. Kasperson, K. Dow, D. Glolding, and J. X. Kasperson. Earth Transformed Program, Clark University, Worcester, MA.

Long, M. H. and J. I. John (1993). *Risk-based emergency response.* ER93 Conference on the Practical Approach to Hazardous Substances Accidents, St. John, New Brunswick. Canada.

Miles, S. and S. Chang (2006). Modeling community recovery from earthquakes. *Eathquake Spectra* 22(2) 439–458.

Mileti, D. (1999). Designing disasters: determining our future vulnerability. *Natural Hazards Observer* 22(1)1–3. http://www.colorado.edu/hazards/o/septo97/septo97a.htm#Designing.

Morrow, B. H. (1999). Identifying and mapping community vulnerability. *Disasters* 23(1):1–18.

NOAA Coastal Services Center (2007). "Case Study." From NOAA Community Vulnerability Assessment Tool. Available on-line: www.csc.noaa.gov/products/nchaz/htm/methov.htm.

NOAA Coastal Services Center (2007). "Vulnerability Assessment Tutorial." From NOAA Community Vulnerability Assessment Tool. Available on-line: www.csc.noaa.gov/products/nchaz/htm/methov.htm.

NRT (1987). *NRT-1: Hazardous Materials Emergency Planning Guide.* National Response Team (NRT) of the National Oil and Hazardous Substances Contingency Plan, Washington, DC.

Pearce, L. (2000). An Integrated Approach for Community Hazard, Impact, Risk and Vulnerability Analysis: HIRV. Doctoral Dissertation. The University of British Columbia.

Preuss, J. and G. T. Bebenstreit (1991). *Integrated Hazard Assessment for Coastal Community: Grays Hargbor.* U.S. Geological Survey Open File Report 91-441-M.

Quarantelli, E. (1998). Perspectives on the Question. Routledge: New York.

Steinberg, T. (2000). *Acts of God: The Unnatural History of Natural Disasters in America.* Oxford University Press, New York.

Stringer, E. (1999). Action Research, 2nd ed. Sage Publication: London.

Tierney, K. and M. Bruneau (2007). Conceptualizing and measuring resilience: a key to disaster loss reduction. *TR News* 250:14–17.

Tobin, G. and B. Montz (1997). *Natural Hazards.* New York: Guilford, New York, 5–15.

Turner, B. L., II, R. E. Kasperson, P. A. Matson, J. J. McCarthy, R. W. Corell, L. Christensen, N. Eckley, J. X. Kasperson, A. Luers, M. L. Martello, C. Polsky, A. Pulsipher, and A. Schiller (2003). A framework for vulnerability analysis in sustainability science. *Proceeding National Academy of Sciences* 100(4):8074–8079.

United Nations, Department of Humanitarian Affairs (1992). *Internationally Agreed Glossary of Basic Terms Related to Disaster Management.* (DNA/93/36). United Nations, Geneva.

U.S. EPA (1986). *Emergency Planning Community Right to Know Act* EPCRA, 42 U.S.C. Ann. Section 11001 et seq. U.S. Environmental Protection Agency, Washington, DC.

U.S. EPA (1987). *Unfinished Business.* U.S. Environmental Protection Agency, Washington, DC.

U.S. EPA (1983). *The Integrated Emergency Management System: Process Overview.* CPG 1-100. U.S. Environmental Protection Agency, U.S. Government Printing Office, Washington, DC.

Waugh, W. (1999). Living with Hazards, Dealing with Disasters: An introduction to emergency management. M. E. Sharp: Armonk, NY.

Weichselgartner, J. (2001). Disaster mitigation: the concept of vulnerability revisited. *Disaster Prevention and Management* 10(2):85–94.

White, G. F., Kates, R. W., and Burton, I. (2001). Knowing better and losing even more: the use of knowledge in hazards management *Global Environmental Change, Part B: Environmental Hazards* 3(3–4):81–92.

Chapter 2

Hazard Identification

Objectives

The study of this chapter will enable you to:

1. Define the hazard identification process.
2. Clarify who might be involved in the hazard identification process.
3. Explain the nature of a community profile and how it fits within the hazards analysis process.
4. Explain what mapping community assets and hazards contributes to hazard identification.
5. Explain the problem-solving process and its role in hazards analysis.

Key Terms

100-year flood
Base map
Critical facilities
Hazard profile
Primary and secondary effects
Secondary hazards
Urban and nonurban communities
Community profile
Resilience

Issue

How can public, private, and nonprofit organizations characterize hazards affecting their community and use that information to reduce the adverse impacts of disasters?

Hazard risk zones may be identified in local communities and may surface vulnerabilities that the public has not understood. How do you profile hazards in a community without causing alarm? What information should be shared with the community?

Introduction

Emergency preparedness, response, recovery, and mitigation are based on a careful and comprehensive hazards analysis process. Tierney et al. (2001) note that emergency preparedness and operations plans must be based on a careful examination and understanding of the types of hazards that the community or organization faces. An assessment of a community's social-cultural, economic, and environmental capital is the initial step of this effort. Community assessment and hazard identification is thus a critical part of the emergency preparedness process and forms the basis for a risk analysis and the determination of community adjustments to reduce vulnerabilities (U.S. EPA 1987). These adjustments are accomplished through intentional problem solving throughout the hazards analysis process. Through the problem-solving process we are able to see issues more clearly by looking beyond just the symptoms of a problem and acknowledge that our organizations must address problems that make our communities more vulnerable to disasters. Deyle et al. (1998) stress that community decision makers must understand how hazards could affect their community and form a basis for strategic risk management strategies and hazard mitigation decisions. They point out that hazard identification is the foundation for many decisions associated with emergency management. Deyle et al. stress that it is an essential foundation for emergency management, upon which a hazards analysis is based (1998).

The U.S. Environmental Protection Agency (U.S. EPA 1987) stresses the importance of hazards identification within the hazards analysis process. Hazards identification is the initial step in the hazards analysis process, includes a description of the community, and provides an analysis through environmental modeling of the nature of the threat. Although the EPA places an emphasis on chemical hazards, it does stress the importance of examining both short-term and long-term threats, community engagement, a broad-based multidimensional planning team, and an orientation to action. A key element in the hazards identification process is gathering data about hazards and the inclusion of public, private, and not-for-profit partners in the community.

This chapter addresses the first part of hazards identification by examining the nature of the community and describing social-cultural, economic, and environmental assets. Chapter 3 will build on this by examining environmental modeling and how we can use this tool to characterize the nature of hazards. Characterizing hazards through environmental modeling allows us to identify what problems may be present that can be addressed to reduce vulnerability. A critical assumption within this initial step of hazard identification is that effective response plans must address the unique nature of hazards in a community. Response strategies should be based on the nature of the hazard event; fires, floods, earthquakes, and windstorms present very different challenges to responders. Further, the nature of the community including its demographics, geography, infrastructure, and critical resources all impact how a community responds and its capacity to recover from a disaster.

Critical Thinking: Hazard identification is an ongoing process that enables communities to establish a culture of hazards risk management and hazard mitigation. It acknowledges that natural hazards may reveal more fundamental weaknesses in local systems that must be addressed to truly ensure community resiliency. What types of information are critical in understanding community assets and resources?

Hazard Identification Process

Hazard identification involves the process of describing the hazard in its local context and provides a description and historical background of potential environmental hazards that could impact the community. This process results in a clarification of the magnitude of a hazard that may pose a threat to the human, built, or natural environments. Comprehensive historic data on all hazards are critical in understanding what hazards have impacted a community in the past and their likelihood for the future. The hazard identification process thus includes an examination of past disasters and the potential for future disasters within the community. We stress that even an insignificant hazard could trigger a much larger secondary hazard.

It should be noted that hazards are part of our lives, and all communities face hazards of many types. Smith (2004) notes that disasters have natural, economic, and human impacts including loss of life, the destruction of homes, businesses, and critical infrastructures, and damage to sensitive natural areas such as wetlands or water bodies. Disasters result from the interface between natural, the built, and human systems. Clarifying the nature and extent of this interface allows us to determine where the interactions are constructive and when loss or damage might result. Understanding these interactions forms the basis for organizational policy decision making and hazard risk management.

Smith (2004) notes that natural disasters result from the interface between the natural geophysical systems and human systems, both constructed and personal (Figure 2.1). Our capacity to minimize adverse effects of disasters depends on our human adaptation to natural events, including our building codes, land use regulation, and the design of our critical infrastructure. Our resiliency or capacity to

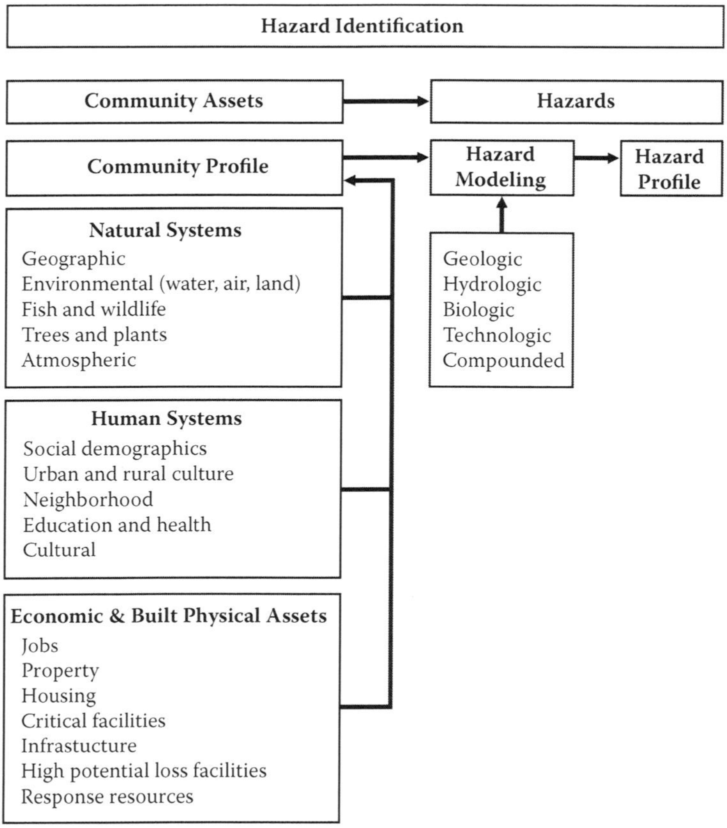

Figure 2.1 The linkages between community assets and hazards.

withstand or to recover from a disaster is influenced by human adaptive actions. We are not helpless objects, but can take steps to protect our social, economic, and natural systems from harm. Burton et al. (1978) stress that the environment is neither benign nor hostile. It is in fact neutral, for human action within a hazard zone establishes our vulnerability. Therefore, we must examine our natural, human, economic, and constructed systems to fully understand what actions may be taken to reduce our vulnerability and enhance our resilience to natural hazards. Fundamental to this process is an identification of the hazards that face our communities and organizations. The United Nations stresses the importance of understanding hazards and that slowly evolving hazards such as drought and environmental degradation should be included in our analysis (2002).

Hazard identification clarifies natural and human-caused events that threaten a community. This process results in information that reveals a community's capacity to deal with a disaster. It provides an opportunity for us to identify the physical characteristics of buildings, the social characteristics of our community, and local response capabilities. Hazard identification may be directly used in preparedness activities by clarifying hazard zones for response, but Deyle et al. (1998) note that it can be used for establishing setbacks and zoning classifications. The comprehensive identification of hazards can support hazard risk management policies and programs and determine benefits of alternative policies to reduce our vulnerability. Hazards analysis thus becomes more of a management tool rather than just one for preparedness.

Organizing a Hazard Identification Team

One does not begin the process of identifying hazards alone; it is done with others from the community. A broad-based representative group such as a local emergency planning community provides a sound basis for characterizing the community and examining what hazards might affect the community. Team members who have diverse technical skills bring different perspectives on just what could happen to the community. Those who have a background in engineering, building sciences, public safety, emergency medicine, health care, education, or transportation are just a few areas that contribute to good when dealing with potential disasters and their impacts.

It would be helpful to form smaller groups within the committee that can concentrate on one or more of the hazards that could affect the community. Doing so enhances our capacity to identify all hazards that could impact the community. For organizations that undertake the hazard analysis process, a diverse team is just as essential. School districts that want to reduce their vulnerability need teachers, administrators, building maintenance staff, transportation, food preparation, utilities, and security on their hazards analysis team. These teams should also be organized in a cooperative way that allows for information sharing between groups, thus reducing redundancy of effort. For large organizations with numerous sites, such as banks, restaurants, or local school districts, there should be consistency among all teams regarding the reporting of findings, the collection of data, and the documentation of sources.

Creating a Community Profile

The goal of the hazard identification and characterization process is to describe hazards that could impact the community or organization. It also provides a basis for future steps in the hazards analysis process. In order to do this, we need to first look at our community and to determine just what is at stake in our community from our people, our economy, the environment, and building and infrastructure. When creating a hazards profile, we categorize hazards into groups by type. However, a hazard in one category may result in a secondary hazard included another category. Heavy rains could cause flooding and lead to chemical spills or an avalanche. The sequence that a hazard evolves and the nature and extent of the resulting disaster will influence how we perceive the threat of the hazard to the community. The division of hazards into these categories helps to provide structure for identifying and characterizing risk to a community or organization and helps us to understand the impacts of hazards.

Social, Physical, and Environmental Assets

The first step in creating a community or organizational profile is to examine its assets. We examine social, economic, and natural parts of the community and their spatial dimensions which could either increase risk or support effective emergency response, mitigation, planning, and recovery activities. There are three principal components of profiling community assets which have been adapted from FEMA guidance, including environmental natural assets, social environment, and the built and economic environment (2001).

A. Environmental or natural assets (geography; land cover; topography; slope; aspect; lakes; streams, lakes and waterways; wetlands; watersheds; soils; fault lines; wildland/urban fire interface; forests; and coastal dunes): These assets may provide natural capital by giving a community beautiful landscapes, resources for recreation, natural flood protection areas, or natural transportation routes for commerce.
B. Social environment: This includes the capacity of members of the community to anticipate, cope with, resist, and recovery from the impacts of disasters. It includes social, cultural, and heritage capital. The level of education and their availability to join the workforce, population distribution, population concentrations, degree of poverty, the nature of volunteerism and community engagement, crime rates, educational attainment, availability of health care and critical heath care rates, vulnerable population groups—children, non-English speaking, elderly, and disabled.
C. Built or economic environment: The nature of our economic capital includes the built and physical environments. Building codes and zoning regulations influence the location and physical assets of a structure. The location and capacity of response resources and facilities that have vulnerable clients or residents will also help or limit a community's capacity to deal with a disaster. The nature and distribution of utilities, transportation, or energy resources will also affect response and recovery efforts. Locations of major employers and financial centers are part of this capital. Our economic capital includes:
   1. Property (land use, types of construction, location of homes, location of businesses, location of manufactured homes, zoning map, and building codes)
   2. Critical facilities (fire stations, ambulance locations, police stations/law enforcement, hospitals, schools, senior centers, day care centers, city hall and other public facilities, prisons and jail facilities, and historic and cultural buildings and areas)

3. Infrastructure (utility lines, water lines, sewer lines, gas lines, pipelines, roads and highways, railroads, airports, waterways, port facilities, bridge locations, communication facilities, transit routes, major energy sources, water purification and treatment plants, landfills, dikes and flood protection structures and facilities)
4. High potential loss facilities (nuclear power plant, dams, military installations, industrial sites that manufacture and/or store hazardous materials, and gasoline storage and distribution facilities)
5. Response (ambulance districts, fire districts, emergency first responder districts, early warning systems, emergency operations centers, hazmat equipment, evacuation shelters, and emergency equipment—fire trucks, ambulances, response vehicles, etc.)
6. Commercial and industrial development from financial enterprises to a community's diverse service sector.

Profiling community assets is an initial step in understanding hazards in a community, for in this process we establish a baseline for characterizing human, natural, and built capital. This baseline allows us to see the strengths and weaknesses of our local community. Where we have strong and robust social, economic, and natural assets, we are more likely to respond and recover quickly from a disaster. The capacity of the community to sustain the impact of a disaster with limited damage and recover quickly involves the concept of resilience.

Environmental or Natural Assets

A community's natural environment plays a critical role in defining hazards that could impact a community's hazard mitigation, response, and recovery strategies. Being located adjacent to a river or in a coastal area could increase vulnerability from a hurricane or severe storm but also provide extensive recreational or commercial resources. Communities located in hilly areas could be vulnerable to landslides. Communities located in wooded areas could be vulnerable to wildland fires.

The health and vitality of the community's natural environment is critical to measuring the community's resilience. A healthy and productive natural environment and ecosystem provides protective buffer for communities. Healthy and productive wetlands provide invaluable flood protection to coastal communities by reducing storm surge and offering broad drainage areas in times of flooding. Healthy forests are less vulnerable to catastrophic wildfires and reduce landslide dangers on slopes. Dunes on coastlines provide buffer from storm surges and winter cold fronts. A damaged, stressed, or unhealthy natural environment can reduce a community's protection from wildfires, floods, storm surges and, in some cases, can increase the damaging impacts of a hazard on the community. A community affected by flooding could in turn have indirect secondary impacts throughout the community. A flood could destroy wetlands by covering them with sediment or erosion, turning productive swamps, marshes, and channels into sterile water features. We need to examine the primary effects of hazards and determine if there are any secondary or indirect impacts.

Communities that allow productive wetlands to be used for development reduce the natural flood protection and increase the risk of flooding. Forest land that is not properly maintained, or where human development is unchecked, can result in wildfires that destroy valuable forestland. Clear-cutting forests on the sides of hills and mountains increases the chances of landslides. Destruction of coastal dunes reduces vital natural protection to coastal communities from storm surges.

Understanding the direct link between a sustainable and resilient natural environment and a community's vulnerability to hazards is critical to developing effective strategies to reduce economic, social, or environmental costs from disasters. Conducting an inventory of the features of a community's natural environment is an important step in assessing a community's assets and in determining how the community is broken up in terms of geography and how the geographic landscape could influence flooding, landslides, or fires.

The recent report on climate change by the Intergovernmental Panel on Climate Change stresses that greenhouse gas emissions at or above current rates will very likely result in further warming (IPCC, 2007). This may induce many changes in the global climate system during the twenty-first century that are larger than those observed during the twentieth century. This report suggests that human actions are having unintentional consequences, and action must be taken now to avert a disaster on a global scale. Our failure to limit excess carbon dioxide emissions, especially from the use of fossil fuels, and methane and nitrous oxide will have adverse consequences in the form of increases in global average air and ocean temperatures, widespread melting of snow and ice, and rising global average sea level.

Global climate change has had a significant impact on the frequency and severity of natural disaster events such as hurricanes, floods, tropical cyclones, drought and extreme heat, and fires in recent years (Abramovitz 2001; IPCC 2007). Unprecedented rainfall in some areas of the world in recent years has resulted in larger than usual flooding events. Hurricane Mitch and Hurricane Andrew both attained Category Five status and resulted in extensive loss of life and damage when they made landfall. Severe droughts have occurred or are ongoing in many areas of the world, including the Southwestern and Southeastern United States.

Whatever the nature of global climate change, one fact is that there is an increase in frequency and severity of natural disaster events around the world. The IPCC notes that extreme events are very likely to change in magnitude and frequency with global warming. Extended warmer periods are likely to cause increases in water use and evaporative losses and duration of droughts (IPCC 2007). Coastal communities worldwide will be the first to feel the effects of global warming. Unfortunately, low-lying coasts already suffering elevated vulnerabilities due to subsidence and erosion will be increasingly threatened by rising sea levels and possibly more frequent hurricanes.

Social Environment

The social environment in a community is represented by its demographic profile. This profile will help emergency planners to determine what segments of the community's population could be impacted by a hazard event. It is essential that local decision makers know which segments of the community or social enclaves will be affected by a single or multiple hazard. What is the community's vulnerability by age, education, income, ethnicity, degree of home ownership or renters, mode of transportation, special-needs residents, single-parent heads of household? Does the community have experience in dealing with hazards?

We learned from Hurricane Katrina in 2005 that low-income renters living in areas that experienced high flood levels for over three weeks needed more public assistance and services to recover. Neighborhoods that experienced extensive damage have recovered more quickly because of access to flood insurance or other family financial resources. Those without flood insurance either could not rebuild or were dependent on additional public support for rebuilding. Areas of the community that are far from major transportation routes, commercial districts, or less damaged areas have had limited rebuilding and recovery following Katrina (Pine 2007).

Neighborhoods that have high percentages of elderly, dependent on public assistance, may have fewer financial resources to bring to the recovery process. Renters may have less financial investment in a neighborhood but have great attachment to the area and be less willing to relocate when disaster strikes. The centennial census provides information on how long residents have lived in an area and if they are homeowners or renters. Additional information on local residents may be collected from churches, neighborhood or social organizations, clubs, fraternal groups, relief organizations, public workshops, and meetings with local officials. Questions may be addressed such as: How many people are in a neighborhood or sector? Which subdivisions are located in high risk parts of a community? Where do people work, recreate, or gather for civic events?

Built Environment

The built environment is a part of the economic capital of a community. It includes residential, commercial, and industrial assets as well as critical sites such as schools and colleges, hospitals and nursing homes, emergency services, day care centers, criminal justice sites, museums and other cultural resources, utility right of ways or sites, and transportation routes and sites from airports, ports, rail yards, or bridges. The built environment also includes political assets such as local building codes and zoning regulations, as well as the extent of local flood control programs and structures.

Many economic activities could be affected by disasters. Major transportation routes can be disrupted, limiting supplies needed for commercial enterprises or manufacturers. Shippers are unable to provide needed materials or carry end products, thus limiting the capacity of the enterprise to successfully carry out their mission. For many businesses, nonprofits, and government agencies, critical fuel supplies, food, and other organizational inputs may be blocked. For some communities, transportation systems (docks, railroads, or bridges) are located near the river to expedite commerce. Although they may have been constructed to a level above anticipated flood levels, flood characteristics change over time as a result of changes in the landscape and land use. New hydrologic modeling might show that higher flood peaks might occur sooner and more frequently as urbanization covers open space with impervious surfaces (roofs, streets, parking lots, etc.).

Infrastructure

The loss of critical infrastructure can cause major direct and secondary disaster impacts in a community, from the destruction of a bridge that disrupts movement of resources and citizens in a community, to a major failure at a power station during winter months where electricity is a life-or-death factor. Awareness of those infrastructure components, and the mechanisms by which they affect the community both geographically and physically, is a major step toward successful management of hazards.

Critical Facilities

Critical response resources such as police stations, fire and rescue facilities, hospitals, shelters, schools, nursing homes, and other structures in the community are vital parts of the built environment. Information on critical facilities may be obtained from the administrating agencies, boards (drainage, levee, hospital), commissions (planning), departments (public works), or institutions (university departments) at the main office or the building manager, the state office of emergency preparedness, tax assessors, financial institutions, and state/federal agencies. If the community has a hazard mitigation plan, the critical facilities are usually identified in that plan. Taking it to the next step, you can analyze community vulnerability by collecting data on building construction type and quality, age, size, footprint, elevation, building capacity, presence of auxiliary power, and potential evacuation routes. We would also examine the exposure of these critical facilities, including the number, types, qualities, and monetary values of various types of property or infrastructure (Schwab et al. 1998).

Economic Activities

Unemployment following a disaster can cause extensive disruption in a community. Following Hurricane Katrina, some colleges including the University of New Orleans and the Louisiana State University Health Sciences Center in the New Orleans area were able to avoid employee layoffs and to shift educational services to Web-based instructional formats and carry on their educational activities. For the medical school, they simply moved their classes outside the disaster area. Internet servers had been relocated to available sites and thus could support adjustments in educational operations. Many schools and colleges were not so fortunate and had no plans for long-term interruption of services. Organizations such as faith-based nonprofit day care centers, health centers, and other service providers were simply shut down, resulting in increased unemployment. Today, many of these organizations have made contingency plans to ensure that services are continued following a disaster.

Sources of information on economic activities in a community may be obtained from the chamber of commerce, the state and local departments of economic development, tax assessors, business associations, planners, university departments, and meetings with local officials. Community zoning information may also be used to (1) characterize parts of a community and the level of activity in manufacturing, retail sales, and general commerce, (2) determine the location of residential housing and rental properties, and (3) clarify how many housing units are in a specific area of the community, what types of housing units are present, and if the property is insured.

Mapping Community Assets and Hazards

Mapping a geographic representation of the community as a whole provides a framework for examining the nature and extent of a specific hazard and its potential impacts on the community or an organization. Deyle et al. (1998) stress that a critical component of hazard identification is a hazard map. They indicate areas that are subject to a specific hazard of various intensities and probabilities. Each hazard that threatens a community will affect it differently. For instance, while it can be expected that heavy snow will affect the whole community in a uniform manner, avalanches will only be a problem where there are steep slopes or gullies. Maps show the spatial extent of the community and provide a basis for us to see the interactions between hazards and people, structures, infrastructure, and the environment (Figure 2.2). Many hazard maps are readily available on the Internet or in public agency records such as local emergency preparedness or hazard mitigation plans. Web maps reflecting flood hazards or wildfire risk may be obtained from national agencies. One should note the date that the map was prepared or how often it is updated.

Hazard maps are used in organizations to help clarify the nature and extent of hazards and form the basis for much organizational decision making associated with preparedness, mitigation, response, and recovery. Maps can show specific areas that require additional studies prior to investment, zoning decisions, or the acquisition of a parcel of land.

Displaying information concerning the nature and characteristics of the natural, social, and built environments provides an excellent means of determining how hazards might impact a community in emergency response activities, as well

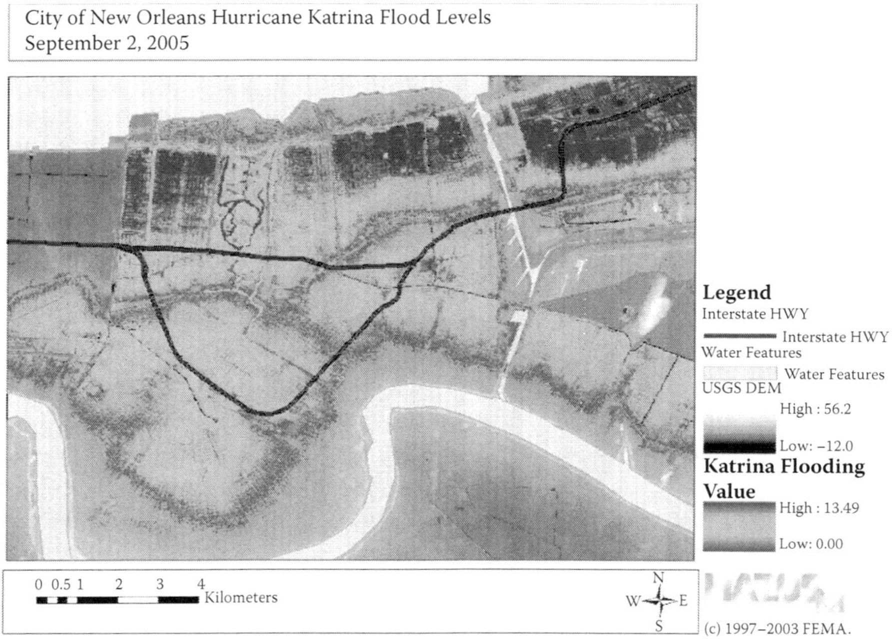

Figure 2.2 (See color insert following page 142.) City of New Orleans, LA, elevation map.

as how specific business, public agency, and nonprofit sectors might be impacted. Communities that are split by major water features, rail lines and large sites such as colleges, manufacturing or commercial enterprises could impair movement of critical resources in a disaster response.

A map is a graphic representation of a physical landscape. It may represent social, natural, or built features, such as transportation networks or political boundaries, or provide information on the number of residents by age, education, or employment by sector. Most importantly, maps provide a means of communicating information. All maps should include the following: a title, mapped area, the agency or person who prepared the map, when it was created, the sources of data used, symbols, legend showing symbols as displayed in the map, direction, and a distance scale. Maps provide a frame of reference for displaying a thematic overlay (hazard) on a geographic base map. A map may simply show where something is located or provide data as in the case of the number of students in a school or patients in a hospital.

A choropleth map is one that presents information using distinctive color or shading to areas (Figure 2.3). The percentage of homes in a community or renters can be represented by grouping data by an area such as a city, county, or smaller representation such as a U.S. Census Bureau track or block group. Maps may simply show data such as population density, population by age groups or education, or

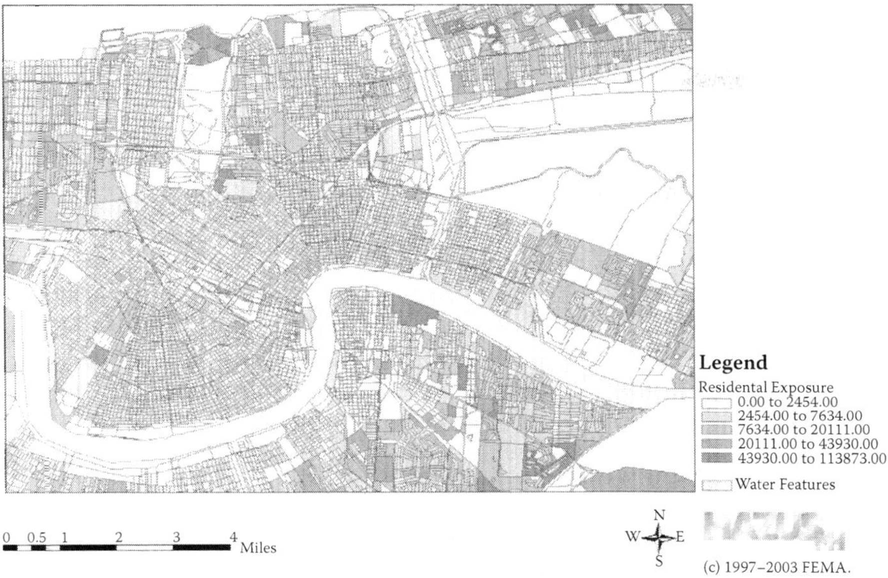

Figure 2.3 (See color insert following page 142.) A choropleth map of New Orleans, LA, showing residential structure values.

even the number of injuries or homes damaged. Maps may give a representation as ratios, proportions, or averages that directly or indirectly involve areas.

To truly compare and analyze the risks of a community, it will be important to have risks represented individually on the base map and together on a single aggregate community risk map. While it will not be impossible to create such a map without a standardized base map, it is much easier, and the resulting product is likely to be more accurate, if standardization is used.

Interdependence of Communities

The inventory of the assets in the community provides data for existing structures, demographics, response capabilities, and environmental conditions. It is also important that information be collected concerning future growth and development plans and how these plans may impact future conditions. It is also critical to recognize that actions taken in one community can impact the vulnerability of surrounding communities to specific hazards. For example, a community that dikes a river will change the natural environment not only in that community but also for communities located downriver, thereby changing the vulnerability to flood events for both communities.

Mutual aid agreements are one example of how communities and states in a region can join forces to ensure that any one community or state will have the needed response capabilities to effectively respond to a disaster event. Mutual aid agreements can include the provision of equipment and staff from one community or state to another in a time of crisis.

Regional growth and development planning is one effective method for ensuring that all potential impacts on neighboring communities can be measured and considered. Understanding that actions taken in one community ripple out to surrounding communities is the first step in preparing comprehensive hazard risk management strategies for a region.

Identifying Community Problems

There is a critical need for information on how disasters are impacting communities. Disaster costs have been rising, and the financial burdens on public, private, and nonprofit organizations as well as for individual citizens has grown to an extent that it threatens the sustainability of many communities (Mileti 1999). Although the number of disasters has not increased, the costs associated with them in terms of property damage, injuries and fatalities, and human suffering have risen dramatically (Abramovitz 2001). The recent costs for response and recovery associated with Hurricane Katrina were estimated in 2006 at $22 billion for direct property losses, economic losses at $8 billion, and emergency assistance estimated at $20 billion, making this the most costly disaster in the United States, and local communities from New Orleans, LA, to Biloxi, MS, will take years to recover (Kates et al. 2006). We have seen that the recovery will take many years, but the devastating social, economic, and environmental impacts motivate us to reexamine how decision making on an individual, family, community, and state level contributes to this continuing rise associated with direct disaster losses. We need to better understand how disasters will impact our communities and how more effective problem solving and decision making can reduce losses. The key is quality information on potential damage impacts, and our past practices are not providing the basis for a systematic examination of problems in our communities.

Hazards analysis along with sound risk management and hazard mitigation strategies provide the initial step in reducing losses. Efforts to reduce costs, injuries, fatalities, and indirect damages unfortunately take time to implement and for us to determine if they are achieving the results that we intended. This is why it is so critical that a hazards analysis, plus risk management and hazard mitigation strategies, be viewed within the context of comprehensive emergency management. Quality planning and mitigation initiatives are the basis for effective emergency response and recovery; they are interconnected and interdependent. Monitoring and evaluating our efforts on an ongoing basis provides a context to determine if our risk management and hazard mitigation initiatives are having desired results.

The fact is that disasters harm people, and it is man who fails to acknowledge the power of nature and take the necessary steps to heed warnings, seek safety, build to withstand storms, and take precautionary actions to avoid losses.

For too long we have focused on understanding hazards and failed to adequately determine how to minimize our exposure to disasters. The gap between hazards and exposure of many types results in problems associated with human, economic, and ecological vulnerability. We need to focus on quality decision making and the adoption of widespread risk management and mitigation initiatives so as to reduce the destructive impacts of disasters. This emphasis on risk assessment and management was given broader recognition by the United Nations International Decade for Natural Disaster Reduction. A worldwide effort was initiated to acknowledge that public policies and individual decisions contribute to disaster losses. Unfortunately, they have not had the desired effects, as demonstrated by the great losses from natural disasters. In fact, most state and local governments continue to conduct limited hazards analysis and do not link the results with problem solving and long-range strategies for risk reduction.

Petak (1985) notes that there are significant challenges to those dealing with hazards, and that problems are rife with complexity and uncertainty, and leaders are not necessarily responsive to them. Further, technical and administrative capacities of organizations are limited in dealing with the complexity of the problems. Clearly, how we go about identifying and dealing with problems that evolve from our analysis of hazards must be carried out in an intentional systematic process. Problem solving always involves risks. To achieve a safer environment, plans and programs must be developed, integrated, and implemented on our ability to understand existing conditions and to predict future consequences.

Problem-Solving Process

Huber (1980) views that organizational decision making lies within the broader problem-solving process. He explains that the first step is to clarify the background upon which the problem situation exists, who is involved, how the situation evolved and, eventually, to develop a clear statement of the problem. This first step may be viewed as background and problem finding. He suggests that in too many cases, managers fail to adequately understand the roots of the problem and how it has evolved. He contends that many managers spend little effort in attempting to focus on the right problem and simply react to the environment in which they exist. From the background information on the situation, one can develop a clear statement of the problem. A problem can be defined as the difference between the current situation and what is desired. Others have expressed it as a gap between where you are now and where you want to be, if you do not know how to get there. Where that difference is great, you have a much larger problem.

The second step is to take the problem statement and determine alternatives for addressing the fundamental issues. This creative part of the problem-solving process is intended to generate many solution options. The result is that one has numerous options in which to pick a strategy for addressing the problem; one option is to combine possible alternatives that have been identified.

The next step is one of choice making and decision making. Huber notes that the decision-making process involves the review of potential solutions to a problem, ranking these options, and then making a choice between them. Decision making then moves to the development of a strategy for implementing the desired solution, and then a follow-up methodology should be established for determining if the selected solution had the results desired.

Critical to dealing with complex problems is the availability of information concerning hazards, impacts, and potential strategies. Wallace and De Balogh (1985) contend that management information systems provide a key resource for effective problem solving.

Huber believes then that decision making lies within a broader problem solving approach that includes problem finding, defining a problem statement, generating options for addressing the problem, determining a strategy for solving the problem, strategy implementation, and finally, monitoring the results of the strategy (Figure 2.4). Did the solution address the problem as desired and are further actions required?

Critical Thinking: Conflicts evolving from problem solving often result from differences in the goals and priorities. It is helpful to articulate clear statements as to the goals of parties, including the analysis of the problem. In addition to our individual goals, our beliefs and values impact what we believe is important and should be obtained. Encouraging those involved in the decision-making process to clarify their goals, beliefs, and values can help move toward a consensus on how to address the problem. How do you see that your values and beliefs might influence your views of problems? How do you approach problems? Do you deal with problems in a systematic way?

We have suggested that an effective hazard analysis involves sound decision making, communication within organizations and with the public, risk management,

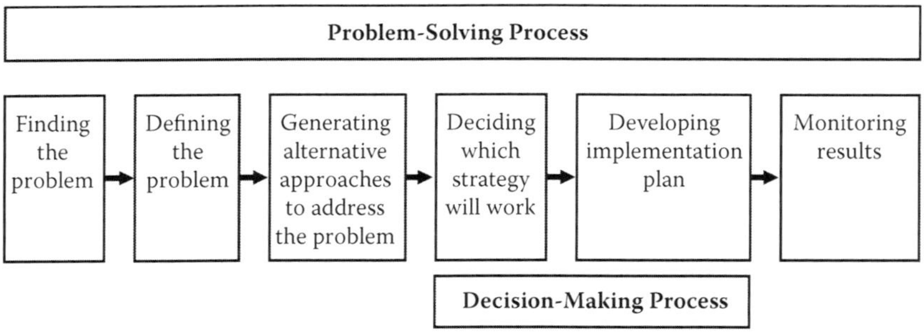

Figure 2.4 The problem-solving process.

and hazard mitigation. These activities compliment and extend the outputs of a hazards analysis and provide the basis for including these outcomes in emergency preparedness, response, and recovery initiatives. The effective management of these activities thus becomes a primary goal. Quarantelli (1996) stresses the need for good management, explaining that this is quite different from just managing. He provides a breakdown of the criteria for "good management" by stressing:

1. There is a difference between agent- and response-generated needs and demands. The first is a tactical need that evolves from a single source; the second is a strategic need that may be identified prior to a disaster and could have long-term implications for the effectiveness of a response. A quality hazards analysis and effective decision making can help identify long-term multihazard response needs.
2. The value of common generic functions is critical to an effective response. Which functions might be included as "generic" should evolve from a comprehensive hazards analysis, including the need for housing, no matter the source of the disaster.
3. A comprehensive hazards analysis can reveal the nature and type of resources that will be needed for almost any disaster. Quality decision making can result in the effective coordination of resources, so that the right people are able to help in an efficient manner.
4. Organizing resources effectively is based on a clear understanding of what is needed, and the hazards analysis provides a sound basis for how a disaster could impact the community or organization. Getting the right people to the right place is not to be taken for granted, but must be managed effectively.
5. Most communication issues that are associated with problems in disaster response relate to communication. Quarantelli stresses that the problems are not in the technology but in what is communicated. The key is that we should stress the content of our communication and not be distracted with just how we deal with others. This is a key insight and stresses the need to communicate clearly the results of our hazards analysis.
6. The need for quality decision making is stressed throughout the emergency management process. Efforts to institutionalize effective decision-making processes, especially in the hazards analysis process, provide a basis for ensuring that a culture of sound decision making is used when there is a disaster.
7. The development of a comprehensive hazards analysis includes many people from throughout an organization and from outside interests. Effective coordination then becomes a fundamental element of good management.
8. Good management also recognizes the contribution that emergency groups and behavior have on successful responses to disasters. An organizational culture that encourages emergency behavior by individuals and groups must be nurtured, and the hazards analysis process is an excellent opportunity to instill this as a positive organization value.

9. Effective management is viewed by Quarantelli to anticipate the needs of individual citizens for information. Working with the media is needed not only in emergency response but especially in communicating the results of a hazards analysis.
10. A quality hazards analysis provides essential information for preparedness and response operations. It should not be viewed as a document to be placed on a shelf, but as a basis for emergency response. Understanding our hazards and clearly articulating the adverse impacts of a disaster provide a solid basis for operations in a disaster.

Quarantelli has suggested that good management does not just happen, but is the result of intentional actions that impact the effectiveness of any response operation. He provides specific recommendations to help us move toward effective management. A comprehensive hazards analysis is a key factor in establishing and maintaining good management.

Problem Solving in an Ill-Structured Environment

Attempting to examine environmental hazards where risk can never be guaranteed and the ultimate solution is never clear presents a situation that is classified as ill-structured (Radford 1981). Decision making that is ill-structured includes the following characteristics:

1. There is a lack of complete information associated with hazards, the environment, and potential strategies that could be considered to reduce the adverse impacts of a disaster.
2. We can quantify the number and frequency of past disasters and their characteristics, but sufficient information to provide a valid estimate of potential disasters is not available. Appropriate measures just are not available to give us a high degree of certainty on the frequency of disasters.
3. Decisions that result from a hazards analysis attempt to address multiple objectives. Unfortunately, these objectives may be in conflict and difficult to quantify. For example, we may want to reduce the damages from a disaster, but local resources are limited, and increased taxes or fees place a greater burden on some than others.
4. Multiple participants that represent alternative positions as well as power bases make decision making complex and ill structured.

A critical strategy for dealing with problems in an ill-structured situation is to ensure that the methodology for examining the problem is clear and stated in advance. This process should provide opportunities for numerous alternative positions to be raised and priorities to be examined. Ensuring representation from alternative positions is critical in ensuring that the outcomes of a decision are supported by all parties.

Decision Traps

Russo and Schoemaker (1990) provide helpful observations for improving the decision-making process by outlining decision traps that pose barriers to effecting resolution of problems.

1. Failing to take the time to adequately consider the problem and prematurely reach a conclusion without putting the issues into some context. The need here is to adequately express what the issues are, who is involved, what is at stake, and potential difficulties that could be encountered if the problem is not addressed.
2. Too often we do not take the time to clearly define the problem and spend too much effort addressing the wrong issues. The key here is to frame the situation so that the fundamental problem is stated in a clear manner. The frame could be influenced by specific goals of the agencies involved, as well as legal or ethical issues.
3. Attempting to understand hazards and their impacts is complex, and we must admit that we do not have all the answers and that we have doubts concerning the potential consequences of some solutions. We need to be clear about the assumptions that we bring to the decision-making process and how our own values, beliefs, and knowledge influence our analysis.
4. We often look for ways to simplify complex situations and, as a result, fail to appreciate the broad nature of the problem or the consequences of our actions. Maintaining an appreciation of the powerful force of the hazard is a constructive position from which to view the problem-solving process. We need to ensure that, where we need additional viewpoints and position or knowledge, we will get it.

Conclusions

Current hazards analysis processes and decision-making approaches tend to put a great deal of power in the hands of technical experts and professional administrators who are not directly accountable to the public. Elected officials must, therefore, be included as representatives of the public and actively engage in the process of exercising value judgments that will lead to agenda setting, resource allocations, staffing, training, and, ultimately, the effective implementation of a program designed to mitigate against, prepare for, respond to, and recover from disasters when and if they should occur.

The determination of an acceptable level of risk is ultimately a policy one that is based on judgment and not just facts. We can attempt to quantify uncertainty, but in the end, we must confront our limits in dealing with natural hazards. Our decisions combine facts and values, and problems require a framework that can facilitate a broad view of the risks associated with floods, earthquakes, wind, or other hazards. Further, our decisions will be exposed in a public arena and require open discussion of our analysis. Our process, therefore, must be broad based and one that deals with complex uncertain conditions. We must create and sustain a culture of quality disaster planning that includes a comprehensive hazards analysis, sound decision making, and both risk management and hazard mitigation strategies that address community and organizational vulnerabilities.

Discussion Questions

If an organization or community has limited resources, how can they prepare a community profile of hazards that could affect them? Would it be a good idea to perform different risk identification methods to identify the hazards in the community? Why or why not?

How does the nature and the health of a community's natural environment influence the social and build systems?

Why is it important to use a single, standardized map or mapping system to display a hazard profile?

What are the relationships between communities within a region, and how do these factors impact hazard risk management planning?

Applications

Using the hazard identification community profile format to your local community, explore the social, economic, natural, and built systems in your community and list the types of assets that are in your area. Using the hazard profile format discussed in this chapter, identify and describe two hazards that are present in your community.

The community profile provides an excellent basis for examining strengths and weaknesses of a community. Using the community profile that you create in the above application:

What do you see as the opportunities presented by the strengths of your community?

What do you see as the weaknesses of the community? What do you see as the problems that result from these weaknesses? What could be the factors that may have caused these problems?

Hazard Web Sites

NationalRelocation (2007). Relocation services—real estate, mortgages, agents, rentals, movers. Community Data: an excellent source for community information. http://www.nationalrelocation.com.

Scorecard (2005).The pollution information site. Environmental scorecard: a community chemical pollution information site. http://www.scorecard.org.

CityData (2008). A very good site for statistics for many communities. http://www.citydata.com.

Sperling, B. (2005). Sperling's BestPlaces. http://www.bestplaces.net.

U.S. Census Bureau (2008). Census Bureau home page. http://www.census.gov.

U.S. Census Bureau (2008). American FactFinder. http://factfinder.census.gov.

U.S. EPA (2008) Surf your watershed. U.S. Environmental Protection Agency. http://cfpub.epa.gov/surf/locate/index.cfm.

ePodunk (2008) ePodunk—profiling 46,000 cities, villages, towns, townships, and podunks across America. ePodunk Inc. http://www.epodunk.com/.

Wikipedia (2008) Main page—Wikipedia, the free encyclopedia. Wikimedia Foundation, Inc. http://en.wikipedia.org/wiki/Main_Page.

Maps

Yahoo! (2008). Yahoo! maps, driving directions, and traffic. http://wwwmaps.yahoo.com.

National Atlas of the United States (2008). National Atlas home page. http://www.nationalatlas.gov.

Google (2008). Google Earth. http://www.googleearth.com.

References

Abramovitz, J. (2001). *Unnatural Disasters.* Worldwatch paper 158. Worldwatch Institute, Washington, DC.

Burton, I., R. W. Kates, and G. F. White (1978). *The Environment as Hazard.* Oxford University Press, New York.

Deyle, R. E., S. P. French, R. B. Olshansky, and R. G. Patterson (1998). Hazard Assessment: The Factual Basis for Planning and Mitigation. In *Cooperating with Nature: Confronting Natural Hazards with Land-Use Planning for Sustainable Communities*, ed. R. J. Burby. Joseph Henry Press, Washington, DC.

FEMA (1997). *Multi-Hazard: Identification and Risk Assessment.* Federal Emergency Management Agency. Washington, DC.

FEMA (2001). *Understanding Your Risks: Identifying Hazards and Estimating Losses.* Federal Emergency Management Agency. Washington, DC.

Huber, G. P. (1980). *Managerial Decision Making.* Scott, Foresman Glenview, IL.

IPCC (2007). Impacts, adaptation and vulnerability: working group II contribution to the Intergovernmental Panel on Climate Change. Intergovernmental Panel on Climate Change, Geneva, Switzerland. http://www.gtp89.dial.pipex.com/chpt.htm.

Kates, R. W., C. E. Colten, S. Laska and S. P. Leatherman (2006). Reconstruction of New Orleans after Hurricane Katrina: a research perspective. *Proceedings of the National Academy of Sciences* 103:14653–14660.

Mileti, D. (1999). Designing disasters: determining our future vulnerability. *Natural Hazards Observer* 22(1)1–3. http://www.colorado.edu/hazards/o/septo97/septo97a.htm#Designing.

Minnesota, State of (2002). *Creating a Community Assessment.* MN Planning, St. Paul.

Monmonier, M. S. (1996). *How to Lie with Maps*, 2nd ed. University of Chicago Press, Chicago, IL.

Petak, W. J. (1985). Emergency management: a challenge for public administration. *Public Administration Review,* Vol. 45, Special Issue, Emergency Management: A Challenge for Public Administration, 3–7.

Pine, J. C. and H. Wilson (2007). Community at risk: an examination of four neighborhoods flooded from Hurricane Katrina in New Orleans. *Journal of Race and Society* 3(1), 7–27.

Quarantelli, E. L. (1996). Ten Criteria for Evaluating the Management of Community Disasters. Disaster Research Center. *PRELIMINARY* PAPER. University of Delaware #241. http://dspace.udel.edu:8080/dspace/handle/19716/649.

Radford, K. J. (1981). *Modern Managerial Decision Making.* Reston Publishing Company, Inc. Reston, VA.

Russo, J. and P. Schoemaker (1990). Decision Traps. Simon and Schuster: New York.

Schwab, J., K. C. Topping, C. C. Eadie, R. E. Deyle, and R. A. Smith (1998). Hazard Identification and Risk Assessment. *Planning for Post-Disaster Recovery and Reconstruction.* Planning Advisory Service Report No. 483/484. American Planning Association, Washington, DC.

Smith, K. (2004). *Environmental Hazards: Assessing Risk and Reducing Disaster.* Routledge Press. London

Tierney, K. J., M. K. Lindell, and R. W. Perry (2001). *Facing the Unexpected: Disaster Preparedness and Response in the United States.* Joseph Henry Press: Washington, DC.

United Nations (2002). *Living with Risk—A Global Review of Disaster Reduction Initiatives.* The Inter-Agency Secretariat of the International Strategy for Disaster Reduction, New York.

U.S. EPA (1987). *Technical Guidance for Hazards Analysis: Emergency Planning for Extremely Hazardous Substances.* United States Environmental Protection Agency, Federal Emergency Management Agency, and U.S. Department of Transportation. Washington DC. http://www.epa.gov/swercepp/p-tech.htm#nrt-1.

Wallace, W. A. and F. De Balogh (1985). Decision support systems for disaster management. *Public Administration Review,* Vol. 45, Special Issue, Emergency Management: A Challenge for Public Administration. 134–146.

Chapter 3

Modeling Natural Environmental Hazards

Objectives

The study of this chapter will enable you to:

1. Clarify the role of environmental hazard models in hazards analysis.
2. Identify the nature and types of environmental models.
3. Explain the criteria one could use in assessing natural hazard models.
4. Explain the advantages and disadvantages of hazard models.
5. Define and discuss the purpose and the elements of a hazard profile.

Key Terms

Areal Locations of Hazardous Atmospheres (ALOHA)
Base flood
Deductive reasoning
Deterministic models
Digital Elevation Model (DEM)
Dynamic models
Experimental design
Flood discharge values

Flood insurance rate maps (FIRMs)
Floodplain
Hazard profile
Hazard risk vulnerability zone
Hazards United States Multi Hazard Flood (HAZUS-MH)
Hydrolic Engineering Center River Analysis System (HEC-RAS)
Hypothesis
Model validity and reliability
Models
River gage stations
Statistical models
Uncertainty

Issue

Environmental hazard models are based on a theoretical framework, assumptions, and a set of interrelated dynamics that can change over time. Users of models must understand the purpose and limitations of hazard models and how they can be used in decision making, public policy, and the development of hazards risk management and hazard mitigation strategies.

Introduction

The Role of Hazard Modeling in Hazards Analysis

The Army Corps of Engineers, the U.S. Environmental Protection Agency (U.S. EPA), the National Oceanographic and Atmospheric Administration (NOAA), the United States Geological Survey (USGS), the Federal Emergency Management Agency (FEMA), the U.S. Department of Defense (DOD), and the Department of Homeland Security (DHS) have utilized hazard modeling and mapping for many years to clarify the nature and extent of tropical cyclones, inland flooding, wind, fire, earthquakes, explosions, radiological and nuclear hazards, landslides, chemical releases, and volcano hazards. The Tennessee Valley Authority (TVA) and the U.S. Army Corps of Engineers (USACE) have also been leaders in the initiative to characterize the nature of hazards using hazard models and maps. Congress authorized the National Flood Insurance Program (NFIP) in 1968 with the enactment of the National Flood Insurance Act, which was administered by the U.S. Department of Housing and Urban Development (HUD) (FEMA 1997). FIRMs were prepared for communities throughout the United States and based on hydrologic modeling for drainage basins. These maps give us clear examples of the use of hazard models in community hazards analysis to reduce community vulnerability. More importantly, they serve as a basis for hazard mitigation and community preparedness programs.

Models are a simplified representation or a physical phenomena (Brimicombe 2003; Drager et al. 1993). In the case of hazards, models simulate the nature and extent of a disaster event. We use models to represent natural events, and for determining how a specific hazard could affect a community. Sophisticated computational models are based on complex mathematical formulas and assumptions. Models are quantitative and attempt to reflect the dynamics of physical, economic, natural, and social processes. Models can be reflected in regression lines predicting an output and based on input variables and mathematical formulas that use complex processes within a computer program.

Chorley and Haggett (1968) suggest that models provide many uses within a scientific context. Models help us to:

- Visualize complex processes and interactions that add to our understanding
- Use models as tools for teaching and learning
- Describe a physical phenomenon or process
- Compare and contrast events, situations or processes
- Collect and manipulate data
- Explore or construct new theories or expand current ones

Most of the environmental hazard models that we use today are based on deductive reasoning. That is, one starts with specific observations of the environment and suggests a theory that is based on a hypothesis. An experimental design is determined and based on real data and results in a predicted outcome or phenomena. The model results either verify an outcome, or the assumptions must be adjusted to correct for an error.

As part of the emergency management and disaster science community today, we are able to take advantage of computer technology advancements to use hazard models, but also interpret their outputs. Many environmental hazard models address a broad range of disasters and run easily on a laptop computer.

Critical Thinking: The key to models is the development of internal staff to set up the model simulations and interpret the results. The models may have grown in capacity to simulate very complex environmental hazards and thus could be beyond the capacity of current professional staff. What may be needed are external resources to help in setting up community hazard model simulations and then assembling a local team to analyze the model results. Scientific support from local universities or consulting organizations could set up models for a jurisdiction and help in adjusting the model inputs for various hazard scenarios. An interdepartmental team could also be assembled from local agencies such as public works, planning, geographic information systems, engineering, public health, and health care and emergency service agencies to explore how the results could impact the community (Pine et al. 2005). What natural hazard models are being used in your community? Who is involved in setting them up and using them?

The HAZUS-MH Flood module distributed by FEMA allows the use of HEC-RAS, which is a riverine modeling program used by the engineering community to describe and simulate inland flooding events. The HEC-RAS model may have been run for a local community as part of a FEMA flood study. FEMA contractors use this widely accepted flood model in preparing adjustments to FIRMs. Obtaining the file from FEMA or the contractor who completed the study allows the local jurisdiction to utilize a well-regarded technical hazard model in a local hazards analysis. Traditionally, FEMA asks local public works, engineering, or planning officials to work with engineering consultants in ensuring that the revisions to FIRMs represent local conditions.

Most models have limitations that impact their use and application. For example, the EPA and the NOAA developed an air dispersion mode. In the initial setup of the model, the user is warned that the ALOHA model should not be used with chemicals that are a mixture of hazardous substances, particulates, or incidents lasting longer than one hour. An analysis of the user documentation stresses that the model provides an approximation of the risk zone or an area that could have property damage, injuries, or fatalities. The user of any environmental hazard model must understand how the assumptions contained within the model affect outputs and how variations of data input could impact results. Errors in data input by the users of hazard models can lead to distortions of the hazard vulnerability zone so that the hazard zone outputs do not reflect the real danger in the simulated hazard. It is critical that data inputs reflect the scenario and the best data available.

We have used USGS digital elevation model (DEM) elevation grid files in hydrological riverine modeling. The grid file is used as a basis for showing where water would flow and, given specific flow rates, just how high water in streams, bayous, and rivers might go. The resolution of the USGS DEM files was expressed in a 30-meter grid. Today we may obtain much higher resolution data using laser technology and establish a 5-meter grid file. The light detection and ranging (LIDAR) files are based on numerous data elevation points and thus provide the basis for determining higher resolution elevations. The difference in the 30-meter and 5-meter resolution is reflected in both the data resolution and in its accuracy. When one compares two DEM files at different resolutions, the lower 30-meter-resolution file is accurate for some spatial representation within the boundary of the grid. The 5-meter higher resolution DEM may have additional values for elevations within the same area. The higher resolution grid DEM may thus show greater variations of contours and elevations simply because more data points were used in constructing the grid DEM files.

Variations of data resolution that is used as input into the model can influence results. Figure 3.1 shows a high-resolution digital elevation model (USGS DEM)

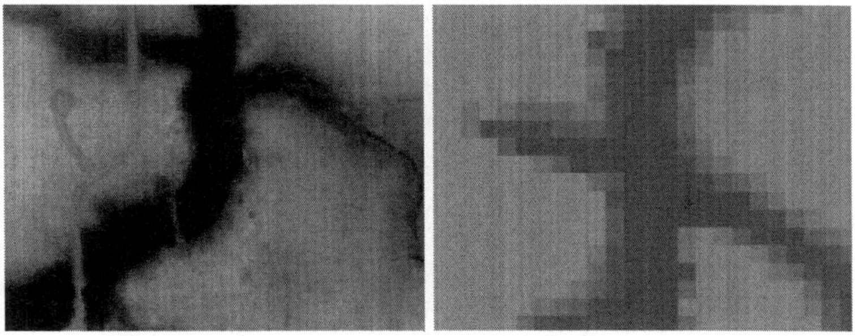

Figure 3.1 DEM files at 30-meter and 6-meter resolutions.

obtained from LIDAR and the older version of the (USGS) DEMs. In the past, most USGS elevation contour data was based on a 30-meter-resolution data format; LIDAR is a new technology that measures the contour of the Earth's surface. The new version of the USGS DEM is formatted as a 6-meter grid. For flood hazards, the higher-resolution LIDAR DEM reveals areas of the landscape that are lower in elevation and could be impacted by flooding. The lower-resolution 30-meter-grid DEMs are not able to show the level of detail in potential flooding as with the higher-resolution LIDAR DEMs. Figure 3.1 provides an illustration of the differences in the two data sets. One can see greater changes in the 6-meter DEM files when compared to the lower-resolution 30-meter files.

Critical Thinking: To see the difference between a 6-meter data set and a 30-meter one, estimate a 30-meter distance and then one that is approximately 6 meters. You are able to see that the 6-meter-resolution data is able to show greater detail in changes in land elevation. Using higher-resolution data allows us to more precisely model simulated hazard events.

Linking GIS and Environmental Models

Brandmeyer and Karimi (2000) established a typology for categorizing how geographic information systems (GIS) and environmental models interface. The most simplistic relationship is one of "one-way data transfer," which allows for a one-way link between the GIS and an environmental model. HAZUS-MH Flood provides an illustration of this type of linkage, where a text file composed of a previous model results from HEC-RAS is linked to HAZUS-MH. The GIS within HAZUS-MH takes the values of elevations within HEC-RAS and determines a depth grid and flood boundary for a specific model run. In this example, if changes are to be made in the scenario, they must be made in HEC-RAS prior to importing the output into HAZUS-MH.

A more complex relationship is described in a loose-coupling type of integration. In this category, there is a two-way interchange between the model and the GIS, allowing for data exchange and change. Processing of environmental data may be made in a GIS using spatial analysis tools, and then the data is moved to the model as a data input.

A shared-coupling design links shared data sets for the GIS and the model (Kara-Zaitri 1996). HAZUS-MH includes a utility to allow the GIS to display residential, commercial, and industrial building data by census block. In addition, a consequence assessment or damage estimate is determined by comparing the census building data with a flood grid, wind field grid, or other type of hazard grid file (coastal flooding or earthquake). Building damage estimates are thus calculated using a common data set of local building inventories.

A joined-coupling design may also be established where both the modeling and GIS use common data sets, but integration occurs in common script language for both the modeling and GIS (Goodchild et al. 1993). Newer versions of hydrological models have been developed so that as environmental conditions change data inputs may be used to revise hazard outcomes as well as GIS displays. The highest level of integration is one where the modeling and GIS are combined in a common user interface and likely on the same computer. Many functions are joined and shared within the programs including data management, spatial data processing, model building and management, model execution, and finally visualization of model outputs in a GIS.

Critical Thinking: Hazards are very complex phenomena and may include inputs such as wind velocity, surface roughness, air temperature, stream flow, and geographic surface features. Physical features impact the effects of natural events and are included in hazard models in the form of mathematical algorithms or formulas. In using an environmental hazard model, it is critical to review how it is constructed and what data are required. Technical documentation is provided for the models such as HAZUS-MH and provides users the necessary information for clarifying how the model was constructed and should be used. Many environmental hazard models provide this critical documentation.

Many models are developed on a national basis and use data sets obtained for communities throughout the United States. The Census Bureau distributes highly accurate data to represent social vulnerability at the neighborhood level. HAZUS-MH uses data obtained from the Census Bureau to determine the number of residential structures, their value, and when they were constructed at the neighborhood level. This data allows hazard models to determine an estimate of the number of people who might be impacted from a flood, earthquake, or wind hazard. Although the information is updated on a ten-year basis, it does provide a good basis for predicting the consequences of disasters (Myer 2004). Myer (2004) showed that residential housing counts and values were very accurate, but that the commercial and industrial building data in HAZUS-MH was not as accurate as the residential data. The model does, however, provide options for editing the building inventory data by users so as to more accurately reflect the built environment in the local community. Unfortunately, local model users must be willing to take the time and expense to tap local building inventory data and make the edits in the database.

Even with the limits of technology, modeling still provides the best estimate of the potential impact of a natural or man-made hazard events. The outputs from models may provide the basis for determining vulnerability zones to floods, landslides, wildfires, earthquakes, or wind hazards and may be used in various emergency response plans and procedures.

Nature and Types of Models

Mathematical models come in different forms such as statistical, dynamic, or combination (statistical and dynamic together). Statistical models are used to predict or forecast future events by utilizing data from the past. These models compare current hazard characteristics with historical data of similar events. Historical records may cover many parts of the continental United States and include data for over 100 years. Note that data collection methods have changed over time, and our understanding of extreme weather or geologic events is far more detailed today than prior to the application of sensitive direct and remote sensing technology.

Dynamic

Dynamic models function differently and use real-time data to forecast extreme climatic events. For example, a dynamic model might take current wind, temperature, pressure, and humidity observations to forecast a specific storm. This type of model is very useful where we have extensive data on the nature of the environment. This is more likely the case for numerous data sources along coastal areas of the United States and water features in inland areas. The use of powerful computers with real-time hazard data collection has led to great improvements in dynamic models.

Combination

Combination models can take advantage of both dynamic and statistical approaches. For areas of the world where precise data measurements are not available, combination models can take a more global perspective and provide good predictions of hazard events on a regional basis.

Deterministic

Deterministic models are based on relationships which can be seen in many environmental applications. For example, a DEM (digital elevation model) provides a description of locations on the Earth's surface as measured by points or contours related to nearby points. We are able to determine the flow of water on the Earth's surface by examining the relationship between contours or points spatially. An interesting dynamic that is seen in this type of deterministic model is that location matters. Tobler (1970) explains that the "first law of geography" is that "everything is related to everything else, but near things are more related than distant things."

Hydrologic tools such as HAZUS-MH utilize DEM files to examine the relationships between land contours, water levels, and potential models. Models such as this are based on well-researched and calculated relationships between land contours, soil types, and land use, as well as water feature characteristics. The data inputs reflecting land contours, soil types, or water feature characteristics are derived from empirically based data inputs. The data inputs reflect specific geographic locations and thus may not suitable for application to new areas. Data is fed into a model, and relationships emerge, usually in the form of rules. As a result, we are able to represent and examine the relationships between very complex dynamic physical processes over a landscape.

Probabilistic

In 1967, the U.S. Water Resources Council (USWRC) published Bulletin 15, *A Uniform Technique for Determining Flood Flow Frequencies* (USWRC 1967; Benson 1967). The techniques used to determine flood flow frequencies were adopted by USWRC for use in all federal planning involving water and related land resources. This bulletin has been updated several times, with the latest version in 1982. Practically all government agencies undertaking floodplain mapping studies use flood flow frequencies as a basis for their efforts (IACWD 1982). Flood flow frequencies (IACWD 1982) from this national initiative are used to determine flood discharges for evaluating flood hazards for the National Flood Insurance Program (NFIP). Flood discharge values are a critical element in preparing Flood Insurance Rate Maps. Corps of Engineers models such as HEC1 utilize data from this data source to calculate flood discharge values.

Statistical probabilistic models such as HEC1 have been used in the National Flood Insurance Program for many years. The HEC FFA model was developed by the Corps of Engineers in 1995 to perform a flood frequency analysis. It performs flood-frequency analysis based on the guidelines delineated in Bulletin 17B, published by the Interagency Advisory Committee on Water Data in 1982 (IACWD 1982). The model estimates flood flows having given recurrence intervals or probabilities; these calculations are needed for floodplain management efforts and the design of hydraulic structures. The program estimates annual peak flows on recurrence intervals from 2 to 500 years. It characterizes the magnitude and frequency of annual peak flows for water features.

Most hazard models determine a risk vulnerability zone for a specific hazard and suggest that individuals in the risk zone could be injured or, even worse, a casualty. Flood models could suggest that residential, commercial, and industrial property could be at risk or vulnerable to flooding if structures are located in an area near a water feature. To determine if specific structures would actually be flooded, additional information is needed about the precise location of the structure, if the building is elevated, and the ground elevation of the structure. If this type of data is not available, then the model would not be able to determine the extent of flooding for a single building in the flood zone. It might flood, or the water might not reach the flood elevation of the structure.

Hazard Models

Some hazard modeling programs, however, do go beyond determining the vulnerability of individuals and property. FEMA and the Defense Threat Reduction Agency (DTRA) collaborated on a multihazard program, Consequence Assessment Tool Set (CATS), that utilizes hazard modeling to clarify the risks associated with earthquakes, tropical cyclones, hazardous material releases, and risks from explosive, radiological, or nuclear hazards. The CATS suite of models displays hazard model outputs in the form of risk zones for use in understanding the potential impacts of disasters, including building damage, injuries, and fatalities. As a result, it can be classified as a consequence assessment tool, rather than showing who might be vulnerable.

Reality Check: The Army Corps of Engineers completed a risk assessment of potential flooding for the City of New Orleans in an effort to show potential flooding in neighborhoods throughout the city (Figure 3.2). The Web-based utility allows a homeowner or business representative a way of identifying the nature and extent of flooding. Check out this example of risk identification and characterization.

HAZUS-MH Model

In 1997, FEMA issued the first release of Hazards United States (HAZUS) for modeling earthquakes in the United States. In January of 2004, FEMA released HAZUS-MH and broadened the types of modeling that could be carried out at the community or regional levels. The most recent release of HAZUS allows for modeling of not only earthquake risks but also riverine and coastal flooding, wind hazards, and releases of hazardous materials using ALOHA (Areal Locations of Hazardous Atmospheres), a dispersion modeling package developed by NOAA and EPA.

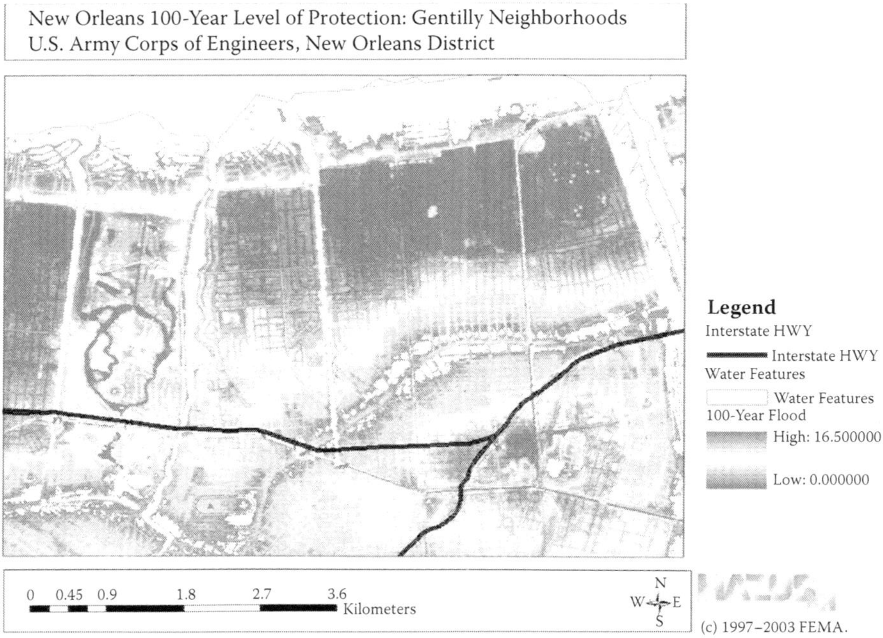

Figure 3.2 (See color insert following page 142.) Flood map of New Orleans—Gentilly neighborhood (http://www.mvn.usace.army.mil/hps/100maps.htm).

The HAZUS-MH mapping and modeling software utilizes the power of geographic information systems (GIS) and hazard modeling to estimate associated social and economic losses as well as characterize the nature and extent of flood, wind, and coastal hazards. HAZUS-MH supports emergency management by enhancing local capacity for determining the potential damage from inland and coastal flooding, hurricane winds, earthquakes, and chemical hazard events (FEMA 2001). Local, state, and federal officials can improve community emergency preparedness, response, recovery, and mitigation activities by enhancing the ability to characterize the economic and social consequences from flood, wind, and coastal hazards (O'Connor and Costa 2003).

Officials at all levels of government have long recognized the need to more accurately estimate the escalating costs associated with natural hazards (FEMA 1997). The Hazard Mitigation Act of 2000 requires that local jurisdictions complete a comprehensive hazards analysis as a part of their hazard mitigation plan in order to qualify for FEMA mitigation funds. HAZUS-MH provides needed tools to estimate the adverse economic impact of flood, wind, and coastal hazards in a community.

HAZUS-MH is just one of the utilities that are available to communities and organizations to characterize risks associated with natural hazards. Allowing local communities and organizations the opportunity to model natural hazards and control the nature of disaster scenarios used in the planning and mitigation process builds modeling capacity at the local government level. Modeling programs for clarifying the nature of natural hazards have long been available to the higher education research community, federal agencies, and their research laboratories associated with USGS, EPA, NOAA, FEMA, NASA, and the Army Corps of Engineers. It was not until 1988 that NOAA and EPA developed and released ALOHA for use by local jurisdictions in emergency planning and response. ALOHA has proven that local communities have the capacity and interest in utilizing hazard modeling in local emergency planning, response, recovery and mitigation activities.

Today, with the use of HAZUS-MH, CATS, and other user-friendly fire, landslide, and volcano modeling programs, local communities and organizations can develop the capacity to use modeling within their organizations rather than remain dependent on engineers and environmental scientists from our research institutions, state or federal agencies, or private consulting companies. Communities have the opportunity to develop in-house capacity for hazard modeling and mapping.

HAZUS-MH Analysis

HAZUS-MH Flood provides basic and advanced analysis for flood hazards and their impacts. The basic analysis uses USGS Digital Elevation Model (DEM) surface grids and discharge frequency values from either the National Flood Frequency Program (Jennings et al. 1994) or, when available, USGS gage stations. The advanced analysis uses either USGS DEM surface grids or higher-resolution DEMs from LIDAR.

Advanced flood modeling in HAZUS-MH utilizes hydraulic analysis from the USGS HEC-RAS (Hydrologic Engineering Centers River Analysis System). As is required for a basic analysis, users conducting an advanced analysis must identify a flood study area and obtain a USGS DEM for the area. A USGS website link within HAZUS-MH provides the connection to obtain a USGS mosaic of 30-meter DEM and 10-meter quads specific to the study area. The mosaic file is in the form of a GRID file and reflects the surface elevations throughout the study region.

Outputs from the advanced analysis using HEC-RAS are the same as the basic analysis; the depth grid, however, is determined from an engineering hydraulic analysis rather than the general statistical discharge estimates reflected in the National Flood Frequency Program or values from the USGS river gage system. Depth is determined for a specific flooding event by comparing the flood elevation along a water feature with the land surface elevations as denoted in the GRID file. Flood elevations for specific cross sections of the water feature are determined using HEC-RAS.

The initial input into a community's hazard mitigation or emergency preparedness program may be from HAZUS-MH basic flood analysis. This type of general analysis renders a foundation for an assessment of the nature and extent of flooding in a study area. The damage calculations reflected in the basic flood analysis help form a general comparison between regions in the study area. This basic analysis establishes a basis for prioritizing future analyses using the advance features of HAZUS-MH and the HEC-RAS. Local jurisdictions may utilize advanced flood analysis capabilities of HAZUS-MH by incorporating previous HEC-RAS into the program. Time constraints are a limiting factor, because setting up each HEC-RAS study area requires geo-referencing the cross sections of peak water elevations. In regard to clearly stated limitations, the HAZUS-MH documentation states that a level 1 "basic analysis" is a generalization of the flood hazard in a local jurisdiction. The hydraulic analysis is not specific to each part of the study area, but is derived from the National Flood Frequency Program. USGS regression equations and gage records are used to determine discharge frequency curves. The depth grid that is the output from the HAZUS-MH basic flood analysis is a much more detailed illustration of flood range depths than what is viewed on a Flood Insurance Rate Map (FIRM). Level 1 analysis is the simplest type of analysis requiring minimum input by the user. However, the flood estimates are crude and are only appropriate for initial loss estimates to determine where more detailed analyses are warranted. Some refer to this type of analysis as "screening." Further studies using the HAZUS-MH advanced level 3 analysis are required for specific decision making at the local level.

Case Study: Data Sources for Flood Modeling

Obtaining accurate data for a local area to characterize a specific hazard is a time-consuming process. FEMA, through the HAZUS-MH program, has collected much of the data for the United States so as to make this process easier. One can do a generalized hazard identification for flood risks using the tools in HAZUS-MH Flood, but additional data is needed for a more precise characterization of flood risks for 100-year events for a local area. HAZUS-MH Flood allows the user to utilize local HEC-RAS modeling files that may have been used by FEMA, the U.S. Army Corps of Engineers, or engineering contractors to characterize local flooding as part of the National Flood Insurance Program. The following examines how data might be obtained at a local level to help clarify factors that could influence flooding in a local area.

Impermeable Surfaces

Characterizing impermeable surfaces in a drainage area can be done using numerous data sources. Land-use data sets that characterize residential, commercial, industrial, agricultural, and open space come in many formats at unfortunately fairly large scales, such as 1:100,000 or 1:250,000 scale. These data sets are unfortunately rather crude when attempting to use them in flood modeling where higher-resolution scales are desired. The availability of high-resolution photos for many communities and regions makes it possible to enhance land-use data sets to account for changes in the use of property from either urban development or changes of use from agricultural to another use.

Topography and Steeply Sloped Drainage Areas (Water Resource Regions and Subregions, Basins and Subbasins, and Watersheds)

Digital Elevation Models (DEM) provided by either the U.S. Geological Survey or other source is the primary source of data for flood modeling that addresses the nature and character of drainage areas. For some parts of the United States, the only source of DEM data is the 30-meter-resolution grid files, while in a few states, efforts to collect and distribute high-resolution 1:6 meter grid elevations using LIDAR technology have been initiated. Having the higher-resolution DEM is especially suitable for characterizing flooding at a local level. The scale of analysis and the source of the data need to be very high resolution, since flooding is very locally driven.

Constrictions

Changes in floodplains can inhibit flood flows, backing up floodwaters onto upstream and adjacent properties. High-resolution air photos can be used by modelers to see if the model output accurately reflects changes in the local landscape.

Obstructions

Bridges, culverts, and other obstructions may be included in the flood modeling program, but if not, certainly should be assessed by those doing the modeling to account for local impediments in the water feature. High-resolution images of the water feature can reveal if there are any obstructions present that might change the flow rate. Access to high-resolution photos is an excellent source of data to check the model outputs.

Debris

Data reflecting the debris in a water feature will not likely exist and will be revealed only in a flooding event. Determining if the local public works agency has scheduled debris removal from water features would be a first place to look for data.

Contamination

Flood water contamination does not directly impact the flow rate of a water feature. It influences the potential harmful impacts of floodwaters in a community.

Local emergency management agencies have hazardous materials inventory data sets for processors (produce, store, use, or transport hazardous materials) of these substances. The data may not be in a geospatial data set but is address specific, and a data set could be made for a local community.

Type of Soil and Saturation

Local weather data is a key in understanding and characterizing soil saturation and should be available for the study region. Local soil types do vary spatially, but the data exists in digital format at a regional and local level.

Velocity

Stream flow data is collected by sensors on many water features in the United States by the USGS in collaboration with other federal agencies and numerous local governmental units. In some cases, nonprofit organizations who have keen interest in environmental issues have worked with USGS to install stream gage sensors. For a gage to be used in flood modeling programs, there needs to be at a minimum ten years of data.

Ground Cover

Trees of various types and vegetation may be available for communities that have sensitive ecological areas. Access to high-resolution photos can be a very helpful source of data for characterizing wetlands from forest by type.

Size of the Drainage Area

The drainage area may be defined by using a DEM of any resolution. GIS programs can be used not only to define which areas flow into water features but also to define the outer boundaries of a basin or subbasin. The DEM may also be used to define the water feature as it moves through the drainage area.

The ideal data set would be one in which the metadata includes a discussion of the accuracy of each object in the data. It should explain the data collection process account for any error, explaining measures of accuracy for elements in the dataset. Uncertainty should be noted explicitly. Unfortunately, the current status of many datasets of GIS does not adequately address these concerns.

The results from HAZUS-MH are formatted for decision makers in several forms. The reports generated by HAZUS-MH are explicit and easy to read. However, the user is required to complete the hydraulic analysis on major water features in the study area and then calculate the economic losses for each of these water features. This program presents information in an orderly arrangement and in a form that assists the decision maker in hazard mitigation.

It is critical that local government officials, business managers, and nonprofit organizations understand the nature and extent of hazards. A geographic information system (GIS) provides a tool for understanding specific risks. Today, many hazard modeling programs are linked to GIS tools to display risk zones or identify population areas and infrastructure (roads, bridges, utilities, or industrial areas) at risk.

Critical Thinking: Using the highest quality data in the model will provide outputs that are a more accurate indication of local damage impacts. Emergency managers and others that use the results of hazard models should discuss the quality of the data with those who are running the model to ensure that the use of the model is consistent with the data used in the program. Given the potential limitations of the data used in a model, current modeling technology allows the user to predict close approximation to the real event. Review a metadata file by going to the following site: http://atlas.lsu.edu/rasterdown.htm. Look for DOQQ Images and select one of the sets of images (they offer three dates to select from). From the Downloader, search for New Orleans East DOQQ. Download one of the four Quads–they will be in an image format. A sample is provided in Figure 3.3. Included with the files is a "Metadata" file. Review it and see what information is included.

Assessing Hazard Models

Hazard models are key tools for understanding potential risks to communities. It must be acknowledged, however, that no model is perfect, for models are simplifications of reality. Our goal is to be able to obtain what is described as a good fit between the model outputs and what can be observed or process validity. In some cases, we are able to test our model by running a simulation of a real environmental hazard. Validity is thus a key element in determining the effectiveness of an environmental hazard model. Model validity is determined by examining if the outputs provide the same results with the same inputs or model reliability (Brimicombe 2003). These criteria are extremely useful for determining which hazard models would be appropriate for our use in characterizing hazards and their impacts. As a beginning, our criteria will include the quality of the model outputs, timeliness of model functioning, and completeness of model results.

Quality

Quality concerns the overall accuracy of the model in describing the nature and extent of a specific risk under specific conditions. A critical element is model validity or determining if the model results accurately represent the potential damaging impacts of the hazard event on the physical environment. Finally, we like to know if the model is truly accurate and often are able to compare a model scenario to real disaster events. Any differences must be understood and included in guidance provided to users of the model.

Quality also is related to the replication of results when similar scenarios are repeated. Does the model give the same results when replicated and when the model is run for large or small areas? Does the model representing an environmental phenomenon give the same results when replicated? We want consistent results in similar situations. This means that one would obtain the same outputs each time the same parameter values are used in the model. An interesting test for an environmental model is to make a slight adjustment to one input variable to determine if the outputs are the same. You want to see how the model reacts to very small changes in model parameters and then see if the results from the model change.

A quality model is one that has extensive documentation and any strengths and limitations explained to potential users. The key here is that model limitations are stated in a clear, straightforward manner, and appropriate warnings are provided to the user when attempting to learn how to use the system.

Today models do not stand alone and are in many cases coupled with geographic information system (GIS) technology. This combines the modeling program with the display characteristics of a GIS. Parks (1993) noted that environmental modeling tools lacked any spatial data handling and manipulation tools as offered by GIS. But it goes further, for GIS today has a role to play on several fronts. It should not be too surprising that hydrological and hydrogeological models were helping to guide the shift from one-dimensional to two-dimensional approaches, given their need to understand the high sensitive spatial configuration and characteristics of the natural landscape (McDonnell 1996). Many files needed for analysis benefit from the ability of GIS to format data sets or change their characteristics so that they may be more easily used by environmental models. For example, in hydrological riverine modeling, a DEM file is invaluable for determining land elevations around water features. These elevations can be used when water levels rise where the water goes along the banks of main channels and tributaries. But in addition, the elevations can demonstrate where flood waters go when flows are constricted by impediments in the channel such as bridges or culverts or just the banks of the water feature. These elevations can show backwater flooding that moves into areas that are not directly along a water feature but are in the end subject to flood waters from the water feature.

Brimicombe (2003) makes an interesting observation in comparing the GIS and environmental modeling communities. He notes that the GIS community has worked to establish data standards and open GIS access across networks, applications, or platforms. In contrast, the modeling community is larger in size and more diverse in representing hazards. We can anticipate that in the future more modeling programs will be linked across networks, applications, and platforms.

A further use of a GIS is that we can use it as a tool to adapt data for use in displaying the model outputs and complete analytical processes to display model results. The GIS can change a DEM grid file so that it may be used in flood modeling. The adapted file is used with channel cross sections that express elevation measures along the banks of a water feature. The TIN, cross sections, and flow conditions all are used to determine the elevation of the water along the water feature. This high-water calculation is then measured against the DEM to demonstrate the anticipated location and depth of flooding along the water feature. Analytical tools built into a GIS are thus a critical element of the modeling process and far more than just a display tool for demonstrating where a hazardous condition will be seen. Brimicombe (2003) notes that early environmental models did not link outputs with GIS, and with these tools we could move from a one-dimensional output to one that would be two- or even three-dimensional. Coupling environmental models with GIS tools and capabilities is a major breakthrough for simulating hazards and their outputs.

Today, higher-resolution data sets such as a DEM result in greater computational demands and thus the need for greater processing power, random access memory (RAM), and file storage capacity. The developments of more robust desktop computers and the availability of super computers to run more complex environmental models has enabled the modeling community to make great strides in simulating our environment. As a result, we are able to model larger ranges, basins, or coastlines and at higher resolutions than ever before.

Critical Thinking: Using the highest quality data in the model will provide outputs that are a more accurate indication of local damage impacts. Emergency managers and others that use the results of hazard models should discuss the quality of the data with those who are running the model to ensure that the use of the model is consistent with the data used in the program. Given the potential limitations of the data used in a model, current modeling technology allows the user to predict close approximation to the real event. What barriers inhibit a full understanding of environmental hazard model outputs by users?

Our ability to understand the complex relationships between model elements is provided by statistical tools that can prioritize or characterize which parameters influence the simulated model outputs. Brimicombe (2003: 165) notes that when we increase the "number of parameters at smaller units, we raise the level of uncertainty in model outputs and make validation of the outputs almost intractable."

A key criticism of many models is that they reflect a static environment and do not reflect changes such as weather conditions. The EPA program ALOHA has provided for many years the option of user input for weather conditions or direct input from weather sensors. As winds change in direction or velocity, or temperatures vary, the model receives the changing conditions from the sensor, models the results, and displays the outcome of the dispersion of hazardous chemicals either in a text format or on a GIS. The real-time dispersion-modeling program is a great asset in an emergency response situation.

A final element of evaluating the quality of a model concerns its potential use by decision makers. When the model is completed, must the results be formatted for use by decision makers? Will decision makers who use the model results be able to easily apply the results for its intended use? Is there too much information or confusing results from the hazard model? Does overload result from the model outputs?

Timeliness

The timeliness criterion is sensitive to concerns that many day-to-day decisions must be made in a limited time frame or are time sensitive. Many environmental models run on desktop or laptop computers and can be run in just a few minutes. Other programs such as the hurricane storm surge models take several hours to run and require the power of a server. Decisions on how to respond to situations must be made quickly. Timely information has several components: Is the information provided when it is needed for decision making? Is the information from the model updated as needed? When conditions change such as more concrete in the drainage basin displacing more water, is information provided as often as needed or at an appropriate frequency?

We have learned in many models that critical variables in the model may have data that is out of date. Many flood studies result in changes in FIRMs (flood insurance rate maps) and are updated as needed. In many cases, flooding changes because of development in the river basin. Figure 3.3 provides an illustration of development in a rapidly growing community. Despite efforts to provide retention ponds in new subdivisions, flooding could occur. The images provide a contrast in a small area of this growing community in South Louisiana prior to the rapid growth following Hurricane Katrina in 2005. Changes in the landscape as reflected in these images should be included in the model.

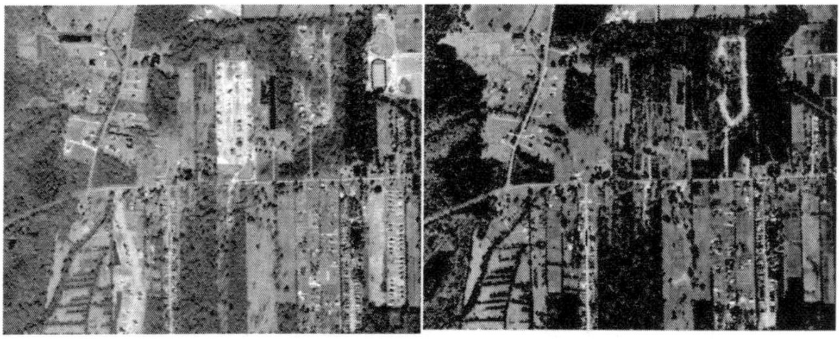

USGS DOQQ 2004
St. Gabriel

USGS DOQQ 1998
St. Gabriel

Figure 3.3 (See color insert following page 142.) Development in a rapidly growing community.

Completeness

The results of the model must be complete to be of value to decision makers. Is the scope of the information sufficient to allow the decision maker to make an accurate assessment of the situation and to arrive at a suitable decision? Does the decision maker have access not only to current information, but also to past history? Are the results of the model presented to the decision maker in a concise form, but with sufficient detail to provide the decision maker with enough depth and breadth for the current situation? Is sufficient relevant information provided to the decision maker without information overload?

The outputs from a model may be complete given the intended use of the results. As an example, a hazard mitigation study prepared for a local jurisdiction should reflect hazards for the area. The basic analysis provided by HAZUS-MH (Figure 3.4) provides sufficient information to allow decision makers to make an accurate assessment of the risks in their jurisdiction and arrive at suitable decisions. The model outputs are adequate for identifying general risk zones, but an advanced analysis using hydraulic modeling results such as HEC-RAS provides more accurate results for decision makers. The more advanced models such as HEC-RAS are suitable for determining base flood elevations and the risk of flooding for specific buildings and infrastructure. This advanced flood modeling capability takes considerable

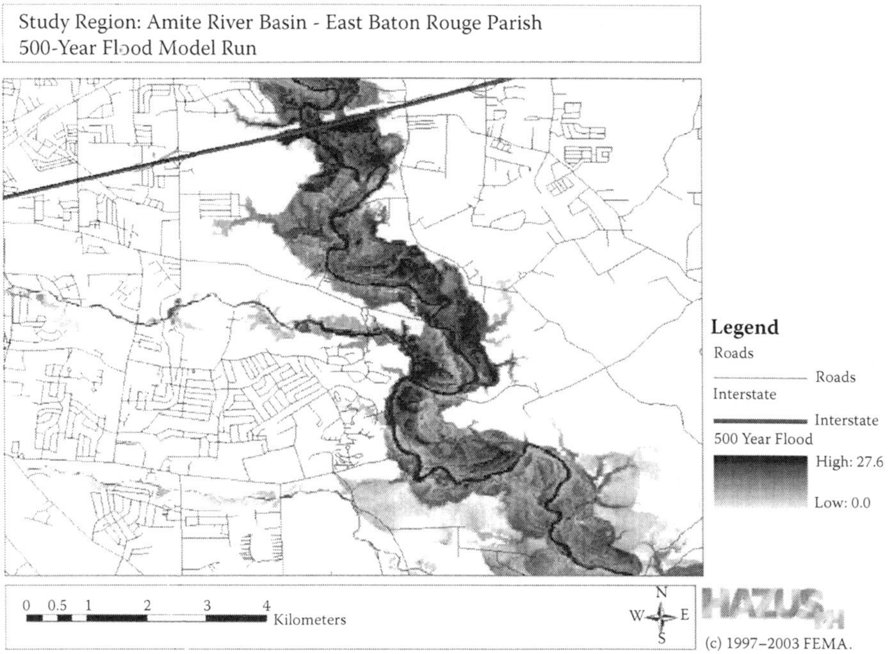

Figure 3.4 Easy-to-read flood estimates from HAZUS-MH.

time and is not possible unless the more advanced flood modeling capability is available to the jurisdiction. A local jurisdiction can set a goal to obtain the detailed hydraulic analysis for their area and input the data into HAZUS-MH.

A balance between conciseness and detail should be determined by the local user where a general analysis may be obtained for understanding the potential for general losses. More detailed information can be obtained with the use of detailed modeling efforts such as HEC-RAS.

Data Accuracy, Resolution, and Availability

Quality data is critical in providing accurate results for planning, mitigation, and decision making. Data issues have become more important today, especially since we are able to obtain higher-quality data for use in models. We do understand that environmental models are a representation of phenomena, and inherent in the process is uncertainty. We just acknowledge that we cannot be 100% correct. Users of our model results must appreciate the inherent errors in our data inputs, calculations, and visual display of the results. Many users from nonspatial disciplines may not appreciate how errors can impact model results. Assuming that our model is correct, we might display the outcome on a USGS Quad sheet, which has a lower spatial resolution than high-resolution images available from private companies or public agencies. Understanding the sources of the data is critical and will be further addressed in the metadata section later in this chapter.

Critical Thinking: Intrinsic uncertainty involves potential error that occurs in our data collection methods. This inherent uncertainty may evolve from the age of the data set, and the area of study may have changed, such as the addition of new housing or commercial developments that alter the landscape, hydrological dynamics, and potential damage impacts to the built environment. This inherent limitation of the model may also result from the capacity of our computer system, including hardware and software. It is thus critical that model users appreciate the purpose and limits of the environmental model and how the output may and should be used.

The date that the data was created could be important to the use of model results. High-resolution images of populated areas that are experiencing rapid growth could provide very different representations of development. The communities that have been experiencing rapid growth could have large subdivisions where just a few years earlier none were present. The difference could be twofold. First, as developers worked the area to be developed, they might have changed the elevation of the landscape so as to alter drainage patterns and influence the rate of drainage into a water feature. Second, roads or commercial areas with paved parking could have also been established and thus increased the level of water draining into a water feature.

A critical part of the record-keeping function in environmental modeling is to record and acknowledge the source of data. This is a customary element of data for any research project and includes documentation on the purpose of the data set, its source, and output data.

Postdisaster modeling to verify the accuracy of a disaster event is a common practice for national agencies. Obtaining data collected by gages at the time of the disaster provides a critical basis for determining if the program functions as the developers stated it would. Federal agencies such as FEMA have sponsored the development of data clearinghouses for major disaster events. See the example of the Katrina and Rita Data Clearinghouse Cooperative as an example of a good data warehouse of disaster data.

A critical component of the above clearinghouse is its ease of access by researchers and agency personnel to find the right data for their use. Since much of the information is raster-based image data, the files are quite large, and one must be sensitive to the time required to move the appropriate data. It should be noted that this site is password protected. Finally, data sources should acknowledge the type of data, and most modelers need to assess the risk of a hazard for a specific geographic area. The data needs to be distributed in an ease-of-use manner.

Critical Thinking: The question of accuracy is also raised by the set of DEM files. A DEM file either 30-meter or 5-meter resolution suggests that the elevations value is the same throughout the spatial area of a single grid cell. It is possible that this is a correct statement; however, it is also quite possible that the ground elevation changes in a grid cell. The high-resolution 5-meter-grid DEM may show changes in land contours that are not seen in a 30-meter resolution DEM. Although this 5-meter DEM is based on many single points in a grid, the actual ground elevation may also vary and have many different elevations within a single grid. When high-resolution 5-meter DEM is compared to a high-resolution photo of the same area, one can see that the 5-meter DEM is a very good representation of the landscape (Figure 3.5). Flood modeling using this DEM would provide very accurate results in these water features.

The clear presentation of hazard risk vulnerability zones is a key part of conveying flood zones, earthquake hazard area, or other hazard risk zones. See Chapter 7 for a more detailed discussion of issues in risk communication.

Advantages and Disadvantages of Hazard Models

Hazard models that are based on comprehensive data sets may provide an accurate representation of a complex environmental dynamic. If this data is up to date, accurate, and in a format that may be used by modelers, then the modeling outputs are more likely to be received in a constructive manner. If users have severe reservations

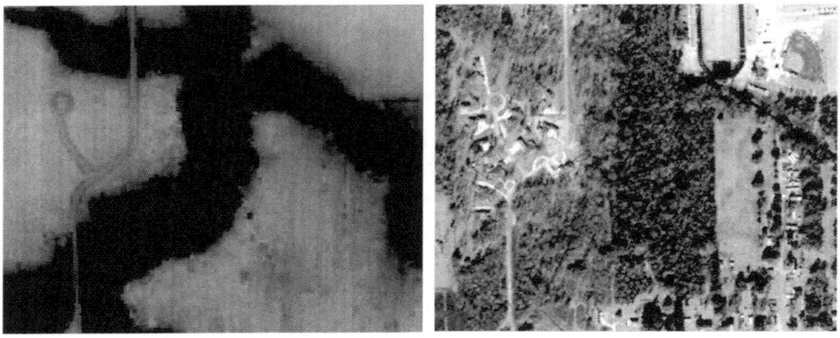

USGS DEM 5-Meter Resolution
St. Gabriel

USGS DOQQ 2004
St. Gabriel

Figure 3.5 (See color insert following page 142.) USGS DEM, 5-meter DEM, and high-resolution image.

concerning data quality that would be used by an environmental hazard model, then they will resist the application of the model in community policy making.

Environmental hazard models are based on a set of assumptions that should be conveyed to the model user and in outputs from the model. These assumptions could involve decisions by model developers that a geographic area mapped was flat, in a rural area, and that data such as weather conditions did not vary over the study zone. Unfortunately, many hazard models include assumptions that are not fully understood by users (Goodchild et al. 1993).

Kirkwood (1994) suggests that environmental hazard models may not provide an accurate representation of the risk to local citizens. He suggests that, when we fail to communicate clearly the nature of environmental risks, public officials and citizens alike can have a false sense of security, or if risk is overestimated, it can consequently cause fear. Unfortunately, environmental models require trained users to interpret the results.

As already noted in this chapter, hazard models that are coupled with GIS can provide more complex information at a higher level of analysis. Risk zones based on different considerations of individual vulnerability can be displayed on a map. For example, a model representing the release of a chemical hazard could show various exposure rates for different populations. Combining maps and graphics gives us an additional tool to provide clear communication to those who want to understand the result of environmental models. Many models may also be able to change data inputs to incorporate environmental and social changes so as to facilitate an accurate understanding of levels of risk or variations of the consequences to disasters.

Environmental data used in modeling hazards should come with an associated metadata file. Metadata provides information about the data set and includes its purpose, an abstract, when it was developed, by whom, any geographic parameters, sponsoring agency, how it is distributed, contact information for anyone who has questions, and use constraints.

Critical Thinking: Agencies have spent considerable time in the development of metadata files that are used in environmental hazard modeling. What is the value of reviewing a metadata file prior to its use in a model?

Hazard Profiles

Once hazards have been identified and mapped in a community, it is possible to create a hazard profile for the community or organization. The hazard profile (Figure 3.7) provides a clear and concise picture of the local context and influence of each hazard, as well as a general description of the hazard for reference uses (FEMA 2001). This information will prove vital in developing a hazard mitigation strategy or emergency response plan and will form a basis for recovery in the event of a disaster. Many hazards have different names, so it is important that the risk statement clearly identifies what is meant by the hazard being profiled. For instance, the hazard "storms" could easily mean windstorms, snowstorms, hurricanes, torrential rainfall, or other hazards. By providing a name of the hazard, and other hazard identifiers that may also be considered in the risk statement, some confusion will likely have been eliminated. There are many measurement and rating mechanisms for hazards that may have changed over time, or may be extremely useful in determining the local context of a hazard. The amount of information provided in these descriptions is determined by those conducting the assessment. However, it is important that both minimum and maximum requirements and expectations are established before the process begins to ensure consistency in reporting, which will alleviate some confusion and ensure that risk perception effects are kept to a minimum.

A hazard profile categorizes the nature of a potential hazard event. FEMA recommends that this type of analysis be initiated as part of the hazard mitigation process so as to fully understand what risks exist to a community. The profile includes:

1. A description of the hazard that could impact a community.
2. The potential magnitude that the hazard could have.
3. The frequency of occurrence.
4. If there is a seasonal pattern to the hazard.
5. The duration of the hazard.
6. The hazard's potential speed of onset.
7. Warning systems that are available.
8. The location and spatial extent of the potential hazard event.

Reiss has provided a similar structure for understanding risks and has encouraged large and small businesses to conduct such an exercise for the organizations (2001).

Type of Hazard

Organize the hazard profile by first identifying the type of hazard that could impact the community or organization. Include both natural and human-caused hazards. Natural hazards include floods, droughts, extreme heat, extreme cold, hurricanes, thunderstorms and lightning, tornadoes, severe snowstorms and blizzards, ice storms, land subsidence, or expansive soils. Natural hazards could also include disease and poisoning. Human-caused hazards include major transportation accidents, hazardous materials spills, widespread power failures, water or sewer failures, telecommunication disruption, computer system destruction, gas line breach, intentional destruction, laboratory accidents involving biological hazards, building collapse, or fires.

Sources of Hazard Information

Research the disaster and emergency history of the community (newspapers, town/city government records, the Internet, public library "local history" section, local historical societies and older members of the community, and local incident reports).

Review existing plans (regional and state transportation, FEMA hazard mitigation plans, environmental, dam, or public works reports, land use plans, capital improvement plans, building codes, land development regulations, and flood ordinances).

Interview locals, risk managers, community leaders, academics, and other municipal and private sector staff who regularly perform risk-management tasks; floodplain managers; public works departments; and engineering, planning and zoning, and transportation departments; fire department; police department; emergency management office staff; and local business personnel. Do a search on FEMA's Web site for state hazard mitigation officers.

Perform site visits to public or private facilities that serve as a known source of risk for the community (chemical processing operations, transportation carriers, and utilities).

Examine a local map of your jurisdiction and note major transportation routes (rail, motor carrier, and water), medical facilities, schools, commercial and industrial locations, apartment complexes, and residential neighborhoods. Examine the relationship between major water features including wetlands and drainage areas (Figure 3.6). What are the critical transportation links in the community? How could they be impacted by a major disaster? How might a single hazard event (flooding) trigger secondary impacts (chemical release)?

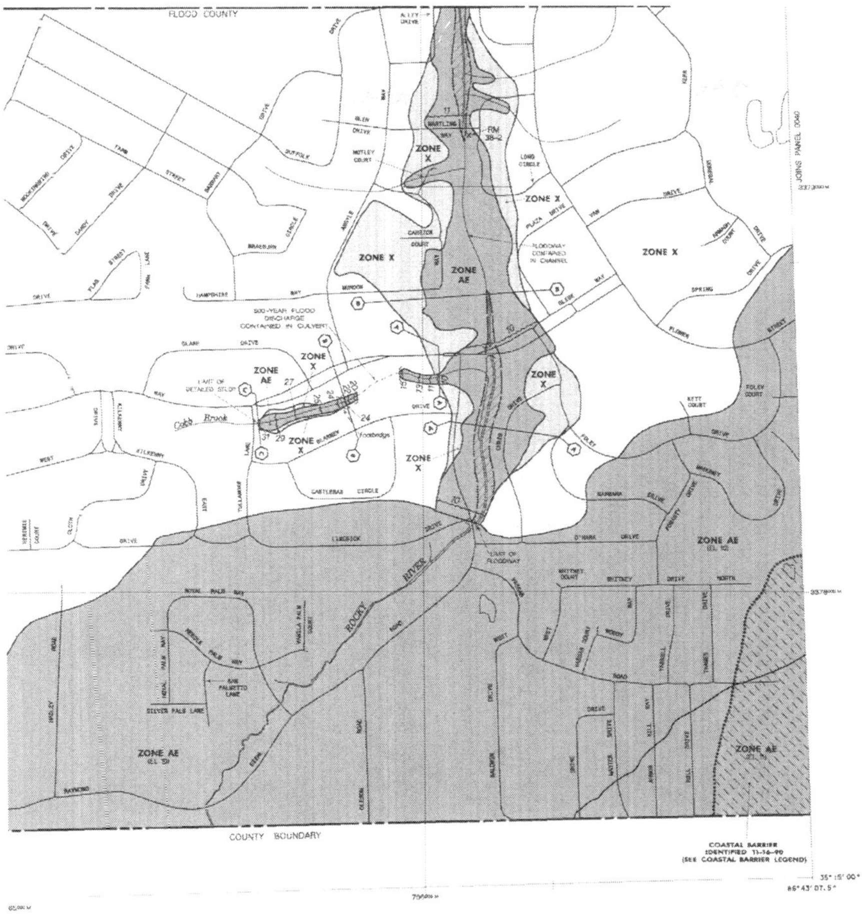

Figure 3.6 FEMA digital FIRM.

Frequency of Occurrence

Historical incidences of the hazard should be displayed in a standardized format, either as a spreadsheet, a chart, or a list. If the hazard is one that happens regularly, it may be better to indicate that fact, and list only the major events that have occurred. This is often true with floods and snowstorms, for example. The predicted frequency of the hazard should be provided as well. Oftentimes mitigation measures, development, or changes in the environment can influence the average annual frequency of a hazard to rise or fall. It may also be helpful to include any comments or reasons why the frequency has changed or is expected to change in the future.

Magnitude

Based upon the hazard maps developed above, this may be a single figure or a range of possibilities. The magnitude and possible intensities will be important in the analysis step, as they will help the team to determine the possible consequences of each hazard and to determine what mitigation measures would be adequate. If a chart detailing the characteristics of a hazard as a function of magnitude or potential intensity was provided in the general description of the hazard, then it is easy to see how the local expected magnitude and potential intensity would be highly useful to the hazards risk management team.

Location and Spatial Extent

For most hazards, the map that was generated above will be sufficient and highly informative in the analysis step of the process. However, if there are some individual areas or regions within the community that need special mention and, likewise, special consideration, this factor should be included as a separate comment in addition to the map. This helps to ensure that those special areas are not overlooked in subsequent processes. This information is also likely to be found on the maps that were developed in the previous step. However, for some hazards there may be special additional comments or facts that need to be added separately from the visual representation provided by the map.

Duration

For hazards that have occurred frequently in the past, it will be possible to give an accurate estimation of the duration of the hazard based upon previous response efforts. However, for disasters that rarely occur or have never occurred, such as a nuclear accident or a specific type of hazardous material spill, it will be necessary to estimate the chance of the incident. These estimations should be based upon the description of the possible hazard, the vulnerability of the community, the response capability of the jurisdiction's emergency response organizations, the response capability provided through mutual assistance agreements with neighboring communities, and anticipated state and federal assistance. This estimate will be measured in days rather than hours or minutes, but will be very useful in subsequent steps that analyze the possible consequences of hazards.

Seasonal Pattern

This is simply a description of the time of year that a hazard is most likely to appear, if such a pattern exists. Knowing seasonal patterns of hazards allows the hazards risk management team to analyze interactions between hazards that could occur simultaneously and devote less time to those that are not likely to occur at the same time. For instance, there is little chance that a town in South Carolina would need to plan for the simultaneous occurrence of a blizzard and a heat wave. While this may seem obvious, this type of information is useful in determining the adequacy of emergency management assets, and for other logistical and planning issues.

Speed of Onset

The speed of onset of a hazard can help planners in the mitigation phase of hazards risk management determine what actions are possible, impossible, and vital given the amount of predisaster time they are likely to have. For a hurricane, there may be several days before the disaster where people could be evacuated and protection measures could be applied to structures and other physical objects, for instance. For a tsunami, there may be 10 minutes to several hours of advance warning, which allows for an immediate evacuation, but little structural work. However, for an earthquake, there is often little or no warning at all, so all preparations must be made with no knowledge of when the next event will occur.

For each of these, the public education and communications systems that can be planned will be drastically different. Warning systems and evacuation plans must reflect the availability or lack of time within which action can be taken. If responders can be readied before the disaster, the speed of response will be increased significantly. For all of these reasons, and many more, it is easy to understand why knowing the speed of onset of a hazard is vital in planning.

Availability of Warnings

This information is indirectly related to the speed of onset of a hazard, but is also independent in some ways. Each hazard is distinct and has certain characteristics that either do or do not lend themselves to prediction through the use of technology or other means. Even hazards that can have a fast onset, such as a volcano, can be predicted with some degree of confidence (though not always), while some hazards with slower onset times, such as biological terrorism, cannot be predicted accurately at all. Lastly, some provide no advance warning at all, such as the case of a chemical accident.

Even if advance knowledge of a disaster is possible, the capabilities of the local warning system further determine the possibility of adequately informing the public about an impending disaster. Local warning systems are more than the physical alarms, sirens, or announcements; they are also the ability of the public to receive, understand, and act upon the warnings they receive. These are all factors that must be considered when determining the availability of warnings.

Hazard Profile Sample for a Tornado

Description of Hazard

A tornado is a rapidly rotating vortex or funnel of air extending groundward from a cumulonimbus cloud. When the lower tip of the vortex touches earth, the tornado becomes a force of destruction. Approximately 1,000 tornadoes are spawned by severe thunderstorms each year.

Tornado damage severity is measured on the Fujita Tornado Scale. The Fujita Scale assigns numerical values based on wind speeds and categorizes tornadoes from F0 to F5. Tornadoes are related to larger vortex formations and therefore often form in convective cells such as thunderstorms or in the right forward quadrant of a hurricane, far from the hurricane's eye. The strength and number of tornadoes are not related to the strength of the hurricane's that generates them. Often, the weakest hurricanes produce the most tornadoes. Large fires can also generate tornadoes.

The path and width of a single tornado generally is less than 0.6 mile. The path length of a single tornado can range from a few hundred meters to dozens of kilometers. A tornado typically moves at speeds between 30 and 125 mph. The average lifespan of a tornado rarely exceeds 30 minutes. (FEMA 1997)

Potential Magnitude

Limited (10% to 25% of area affected).

Frequency of Occurrence

Likely (between 10% and 100% probability in the next year, or at least one chance in the next 10 years). Wayne City lies within a state that experiences an average of 50 to 100 tornadoes per year. The record of events affecting Wayne City during the past 25 years. Please note that, while only one of the four tornadoes directly affected the university, any event occurring within the city had the potential to strike the campus.

Seasonal Pattern

All year, though more common in the spring, summer, and autumn months.

Probable Duration

Between 1 and 30 minutes for the tornado, and up to 24 hours for the response.

Potential Speed of Onset

None to minimal (less than 3 hours).

Identify Existing Warning Mechanisms

The National Weather Service (NWS) issues tornado watches and warnings. The city employs a system of sirens indicating a tornado watch or warning. The university uses a fire alarm system to alert students of any warning. The university also uses the university Web page and university building public announcement systems to warn residents.

Identify Location and Spatial Extent of Potential Event

Entire campus is at risk. The university has identified designated tornado shelters where students are instructed to go during a tornado warning.

For additional information on hazards useful in the development of risk statements and hazard profiles, a list of Web resources is provided at the end of this chapter.

Conclusions

HAZUS-MH provides local government jurisdictions with an exceptional tool in the hazard mitigation process. The ease of use and the quality of the data provided with the program enable the user to generate information on where to expect losses in the community and considerable information on the nature of these losses. Difficulties in using HAZUS-MH will be overcome as local officials become more skilled in utilizing the power of ArcGIS and the riverine hazard model. Training scenarios will likely be prepared to help users anticipate common problems in using the riverine program. Feedback to FEMA on difficulties in using the program will also result in system changes.

One of the biggest concerns by local users is the quality of the data in HAZUS-MH, from the flood frequency values in a geographic area to building counts and their characteristics. Studies to demonstrate the validity of building counts and the economic values associated with the structures will help in assuring local officials of the validity of the data. For jurisdictions where major population shifts have not occurred since 2000, local officials can be more confident in the economic loss analysis provided by HAZUS-MH.

FEMA has created a powerful tool in HAZUS-MH for the assessment of flood losses. The tool allows the user to execute a local analysis in a reasonable period of time and estimate losses to the jurisdiction. It provides a basis for examining the economic impacts of flooding and in establishing hazard mitigation priorities at the local level. By allowing users to establish a multi-county study area, the program facilitates a broader examination of hazard mitigation. Counties can work together to examine flood risks and concentrate more detailed hydraulic advanced analysis in HAZUS-MH to specific areas. As the program gains broad use, local jurisdictions will find help and overcome many technical problems that users are encountering today. HAZUS-MH is thus a powerful hazard mitigation tool. It is a tool that makes a significant contribution in informing local decision makers as to the impacts of flood hazards in their community.

Discussion Questions

Many people contend that detailed scientific results from complex models are often not understood by the public. Why is there such a lack of communication between the scientific community and the public?

Why are some hazard models so difficult to use at the local government level? What is the extent of the use of programs such as HAZUS-MH at the local government level?

Many models combine hazard risk zone outputs with visualization tools such as GIS. How does this tool help in understanding environmental hazard models?

Those who develop and use environmental models stress that the outputs may not reflect real-world conditions. What are the benefits of using models if we can not be completely assured that the results are accurate?

Although many communities and organizations have an all-hazards approach for managing disasters, why is it important to understand how specific hazards could impact a public entity, business, or nonprofit organization?

Applications

Review an NFIP map for your jurisdiction at FEMA's floodsmart.gov Web site, and note when the map was published. For these maps, not all geographic areas are modeled using riverine modeling programs or coastal hurricane storm surge programs. Areas are marked with a reference when the flood zone is based on an environmental hazard model. Other areas are noted in which a flood zone is provided but it was not based on model results. For flood zones that are not based on flood model results, how might the map risk zones be interpreted?

NOAA and the U.S. EPA developed a dispersion modeling program that was intended to be used for emergency response and preparedness activities at the local level. They wanted a program that would be easy to use, provide quick results, but also be reliable and give results that reflected the conditions provided by the user. To download ALOHA, go to the EPA Web site and search for ALOHA. Using the following criteria, evaluate this model for the following:

Does the model provide quality results?
Is the model easy to use and provide timely results?
Are the results of the model complete and usable for decision making in emergency preparedness and response efforts?
Is the data concerning weather conditions, and geographic area, and accurate, at an appropriate resolution, current, and available to the user?

Web Sites

CyArk High Definition Heritage Network. Heritage at Risk Map. http://archive.cyark.org/hazards/?gclid=CM-gzY2EzJECFSUsIgodaRVOzg
National Weather Service Climate Prediction Center. Hazards Assessment Maps. http://www.cpc.ncep.noaa.gov/products/predictions/threats/threats.php

Avalanches

Avalanche Network. http://www.avalanche.org/
http://www.nwac.noaa.gov/nw04012.htm

Coastal Storms

FIRMS also cover coastal regions and provide detailed information concerning coastal flooding or risk from the water-related hazards of coastal storms. The Saffir-Simpson Scale can be used to categorize storms and their wind. Storm surge can be shown on the FIRM and can range from 4 to more than 18 feet above normal sea levels. Inland wind measurements, provided by NOAA, can be used to characterize the maximum sustained surface wind that is possible as storms move inland. Historical data on hurricanes can also be used to characterize future storm events. Data on past storms is provided by NOAA and provides information on tropical cyclone tracks and their associated wind speeds from 1927 until the present day.

FEMA: Hurricane. http://www.fema.gov/hazard/hurricane/index.shtm
National Hurricane Center. http://www.nhc.noaa.gov/
National Hurricane Center. NHC Archive of Hurricane Seasons. http://www.nhc.noaa.gov/pastall.shtml
National Hurricane Center. The inland wind model and the maximum envelope. http://www.nhc.noaa.gov/aboutmeow.html
NOAA. FAQ: hurricanes, typhoons, and tropical cyclones. http://www.aoml.noaa.gov/hrd/tcfaq/G12.html

http://www.nhc.noaa.gov/ (Historical data on hurricanes)
http://www5.ncdc.noaa.gov/

Dam Safety

FEMA. Dam safety. http://www.fema.gov/plan/prevent/damfailure/ndsp.shtm

Drought

Drought monitoring site. http://www.drought.unl.edu/dm/index.html
National Drought Mitigation Center. http://www.drought.unl.edu/index.htm
USGS. Water use in the United States. http://water.usgs.gov/watuse/

Earthquakes

The USGS provides maps detailing peak ground accelerations (PGAs), which measure the potential strength of ground movements that could result from an earthquake. The maps provide the values of the 10% and 2% statistical incidence of such PGAs occurring during the coming 50-year period.

USGS. Welcome to the U.S. Geological Survey Earthquake Hazards Program. http://earthquake.usgs.gov/
http://earthquake.usgs.gov/
http://geohazards.cr.usgs.gov/eq/

National and regional seismic hazard maps: The USGS provides seismic hazard assessments for the United States and areas around the world.

USGS. United States National Seismic Hazard Maps. http://earthquake.usgs.gov/research/hazmaps/
USGS. Seismic hazards maps. http://earthquake.usgs.gov/research/hazmaps/products_data/

Flooding

The FEMA NFIP Flood Insurance Studies (FIS) provide extensive information for local communities concerning potential for flooding. Today these maps are displayed on high-resolution images, which show many local features and reveal vulnerability. These maps show not only water or hydrographic features but also coastlines and extensive risk information from base flood elevations for homes, businesses, and other sites. The maps also show geographic elevation reference marks and, in many cases, elevation contours.

FEMA. FEMA Map Service Center. http://msc.fema.gov/webapp/wcs/stores/servlet/CategoryDisplay?catalogId=10001&storeId=10001&categoryId=12001&langId=&userType=G&type=1
FEMA. FEMA: Maps. http://www.fema.gov/hazard/map/index.shtm

National Flood Insurance Program. Flood Smart for understanding your flooding risk. http://www.floodsmart.gov/floodsmart/
USGS. Natural hazards—floods. http://www.usgs.gov/themes/flood.html
USGS. WaterWatch—maps and graphs of current water resources conditions. http://water.usgs.gov/waterwatch/
USGS. National water conditions. http://water.usgs.gov/nwc/

Heat

NOAA. Heat index forecasts (days 3–7). http://www.hpc.ncep.noaa.gov/heat_index.shtml

Landslides

The best predictor of future landslides is past landslides, because they tend to occur in the same places (FEMA 2001). However, caution should be used because of the complexity of geotechnical factors that determine landslide risk. Prediction requires the expertise of geologists and engineers, who must conduct surveys of soil conditions, slopes, drainage, climate, prevalence of earthquakes and volcanic eruptions, flora cover, erosion, industry-induced vibrations, construction, and other alterations to terrain, among many other factors. The primary elements that help understand landslides include (1) areas that appear to have failed due to landslides, including debris flows and cut-and-fill failures; (2) areas that have the potential for landslides, by correlating some of the principal factors that contribute to landslides—steep slopes, geologic units that lose strength when saturated, and poorly drained rock or soil—with the past distribution of landslides; and (3) areas where landslides have occurred in the past, are likely to occur now, and could occur in the future.

http://landslides.usgs.gov/

Lightning

http://www.nssl.noaa.gov/

Snowstorms

NOAA. NOAA's winter weather Internet references. http://www.noaanews.noaa.gov/stories/s300e.htm

Tornadoes

Communities measure their tornado risk as a factor of the maximum possible wind speed they may experience, should a tornado occur. While tornadoes are highly unpredictable and can occur just about anywhere in the world, there are areas that have historically been more prevalent. This high prevalence exists in the Midwest,

Southeast, and Southwest regions of the United States. Locate the design wind speed for your area and the level of wind resistance that should be factored into the construction of community shelters as determined by the American Society of Civil Engineers (ASCE). Historic tornado data may be reviewed to determine the exact historical frequency of tornadoes in their jurisdiction.

Texas A&M Meteorology Convective Forcast Center. http://www.met.tamu.edu/weather/
The Tornado Project Online: http://www.tornadoproject.com

Tsunamis

Tsunami hazard maps identify inundation zones and can reveal low-lying areas in a community that are at risk. Most maps have been produced at the state level, and considerable local mapping has been done as well. Tsunami maps are available from the West Coast/Alaska Tsunami Warning Center of NOAA/NWS (National Weather Service) Pacific Tsunami Warning Center, Oregon Department of Geology and Mineral Industries, University of Washington, and the Pacific Marine Environmental Laboratory.

NOAA Center for Tsunami Research. http://www.pmel.noaa.gov/tsunami
NOAA. Pacific Marine Environmental Laboratory. NOAA/PMEL is a leader in developing ocean observational systems. http://www.pmel.noaa.gov
University of Washington. Earth and space sciences (geology and geophysics). http://www.geophys.washington.edu
USGS. Natural hazards—tsunamis. http://www.usgs.gov/hazards/tsunamis/

Wildfires

Wildfires are influenced by weather conditions, the availability of fuel, and topography. Fuel availability may be characterized using the Urban Wildland Interface Code (2000) Fuel Model Key. Since fire spreads much faster as slopes become steeper, topography maps can reveal areas that have a high probability of fire. Critical weather elements, such as low relative humidity and high wind, can be helpful in predicting a community's risk from wildfires. This information can be obtained from local and state fire marshals, state foresters, the NWS, or from the NOAA.

National Institute of Standards and Technology. BFRL: Fire Modeling Programs. http://www.bfrl.nist.gov/866/fmabbs.html
NOAA Fire Hazard Mapping System—Satellite Services Division. http://www.ssd.noaa.gov/PS/FIRE/hms.html
NOAA. Storm prediction center fire weather forecasts. http://www.spc.noaa.gov/products/fire_wx/
USFS Wildland Fire Assessment System. Observed fire danger class. http://www.fs.fed.us/land/wfas/fd_class.gif
http://www..usgs.gov/hazards/wildfires

References

Benson, M. A. (1967). Uniform flood-frequency estimating methods for federal agencies. *Water Resources Research* 4(5):891–908.

Brandmeyer, J. E. and H. A. Karim (2000). Coupling methodologies for environmental models. *Environmental Modeling and Software* 15(5):479–488.

Brimicombe, A. (2003). *GIS, Environmental Modeling and Engineering.* Taylor and Francis. New York.

Chorley, R. J. and P. Haggett, eds. (1968). *Socio-Economic Models in Geography.* Methuen and Co., London.

Drager, K. H., G. G. Lovas, J. Wiklund, and H. Soma (1993). Objectives of modeling evacuation from buildings during accidents: some path-model scenarios. *Journal of Contingencies and Crisis Management* 1(4):207–214.

FEMA (1997). *Multi Hazard Identification and Risk Assessment: A Cornerstone of the National Mitigation Strategy.* Federal Emergency Management Agency, *Washington, DC.*

FEMA (2001). *Understanding Your Risks: Identifying Hazards and Estimating Losses.* Federal Emergency Management Agency, Washington, DC.

Goodchild, M. F., B. O. Parks, and L. T. Steyaert (1993). *Environmental Modeling with GIS.* Oxford University Press, New York.

IACWD (1982). Guidelines for Determining Flood Flow Frequency. Bulletin 17B, Hydrology Subcommittee, Interagency Advisory Committee on Water Data, USGS, Reston VA.

Jennings, M., W. Thomas, and H. Riggs (1994). National Summary of U.S. Geological Survey regional regression equations for estimating magnitude and frequency of floods for ungaged sites. U.S. Geological Survey Water Resources Investigation Report. 94–4002, 196.

Kara-Zaitri, C. (1996). Disaster Prevention and Limitation: State of the art; tools and technologies. *Disaster Prevention and Management:* An International Journal 5(1) 30–39.

Kirkwood, A. S. (1994). Why do we worry when scientists say there is no risk? *Disaster Prevention and Management: An International Journal* 3(2):15–22.

McDonnell, R. A. (1996). Including the spatial dimension: using geographical information systems in hydrology. *Progress in Physical Geography* 20:159–177.

Myer, J. (2004). Comparative Analysis between Different Flood Assessment Technologies in HAZUS-MH. M.S. thesis, Louisiana State University and Agricultural and Mechanical College.

O'Connor, J. E. and J. E. Costa (2003). *Large Floods in the United States: Where They Happen and Why.* Reston, VA: U.S. Department of Interior.

Parks, B. O. (1993). The need for integration. In *GIS and Environmental Modeling: Progress in Research Issues,* eds. M. F. Goodchild, B. O. Parks, and L. T. E. Steyaert. Wiley, New York.

Pine, J. C., H. Mashriqui, S. Pedro, and J. Meyer (2005). Hazard mitigation planning utilizing HAZUS-MH flood and wind hazards. *Journal of Emergency Management* 3(2):11–17.

Reiss, C. L. (2001). *Risk Identification and Analysis: A Guide.* Public Entity Research Institute (PERI). Fairfax, VA.

Tobler, W. R. (1970). A computer movie simulating urban growth in the Detroit region. *Economic Geography.* 46(Supplement):234–240.

USWRC (1967). A Uniform Technique for Determining Flood Flow Frequencies. U.S. Water Resource Council: Washington, DC.

Chapter 4

Spatial Analysis

Objectives

The study of this chapter will enable you to:

1. Define spatial analysis and explain how we use this tool in hazards analysis.
2. Explain the type of spatial analysis.
3. Describe how to visualize data using the results of spatial analysis.

Key Terms

Accuracy
Buffering
Choropleth maps
Error
Geospatial data
Hydraulic analysis
Hydrologic data
Hypothesis
Metadata
Precision
Reliability
Spatial analysis
Statistical analysis
Transformations

Issue

What tools are available to examine the spatial and temporal nature of hazards and their impacts?

Introduction

Dr. John Snow unraveled the causes of cholera in the mid-ninth century in London by recording on a map the incidence of cholera. He was able to observe from his map the relationship between a public water pump in the center of the cholera outbreak. Although his use of maps to track the cholera outbreak did not prove the cause, it raised a question as to the relationship between drinking water and the public health outbreak. Stronger evidence was obtained to confirm his contention when the water supply was cut off and the outbreak subsided (Gilbert 1958).

This illustration provides several critical elements of the productive use of spatial analysis in a hazards analysis. First, John Snow collected accurate health data and made accurate georeference placement of this data on a map. He also noted on his map other related items, such as the public water pump. The scale of the map was of a small area within London and provided an appropriate scale in which to test his hypothesis. More importantly, Snow simply used his analysis of spatial data to raise a hypothesis that had to be further studied or tested. He thus limited the use of the information from his analysis and had a sound basis for choosing his data sources and how he would use this data in forming a hypothesis. His methodology was goal directed and determined the scope of his analysis. The key to spatial analysis is clearly stating what we intend to accomplish and determining a methodology that is suitable to achieve the desired results.

Definition of Spatial Analysis

Spatial analysis is a set of tools and methods that are used to examine the relationships between social, cultural, economic, ecological, and constructed phenomena. For our purpose in examining hazards and their impacts, spatial analysis provides a means of understanding the nature of hazards and their social, economic, or ecological impacts. Spatial analysis is the center of how geographic information systems are used in transforming and manipulating geographic data. It provides the methods that are used to support organizational decision making by government agencies, businesses. and nonprofits (Longley et al. 2005). The methods and tools provided by spatial analysis thus give us a means of turning raw data such as what John Snow collected, into useful information. In the case of understanding natural hazards, we can enhance our understanding of the nature of hazards and their impacts by using spatial analysis. The results of our analysis can also help us to better communicate within organizations and with the public. Spatial analysis adds meaning, content, and value to our quest to better understand hazards and their impacts.

Spatial analysis is more than just a fast computer and expensive digital data. It is the formation of a hypothesis and the use of geospatial data in expanding our understanding of how the physical environment interfaces with our social, economic, and natural environments. Statistical methods are used in our analytical methods, but spatial analysis is much more that just crunching numbers. Through spatial analysis, we are able to reveal patterns and processes that otherwise might not have been observed and confirm or disprove our hazard-related hypothesis.

Fisher notes that spatial data analytical techniques perform a variety of functions within a geographic information system (GIS) and are important for the types of questions and concerns that policy makers address in private, public, and nonprofit organizations (1996). He further stresses that using geographic spatial relationships provides a very good framework for understanding the meaning of data within a GIS. Spatial analysis evolved in the early 1960s as part of quantitative geography and the application of statistical processes in examining spatial relationships of points, lines, and area surfaces. A spatial temporal perspective was added to allow us to examine these relationships over time.

Geospatial Data Set

Geospatial data relating to hazards comes in many forms and enables us to characterize both the nature and extent of the hazard event and the many elements that help shape or characterize the hazard.

For flooding events, we need high-resolution elevation data to describe the broad geographic area that makes up a river basin, subbasin, or drainage area. Further, we need to characterize the size and shape of water features that make up the river basin and characterize over time the amount of water at any given time in the water feature (discharge). Other factors that influence flooding in a river basin include the amount of impermeable services (paved roads or parking lots and residential structures, commercial buildings, or industrial sites). Flooding threats in a river basin may change over time if property near water features is changed from a natural landscape to one that has new subdivisions, commercial development, or major changes in roads or parking lots. Rain may flow more quickly into a water feather as a result of changes in the development of the landscape.

Riverine flood models use discharge values, soil types, and land-use and elevation data in characterizing flooding events in a river basin or drainage area. These flood-modeling programs utilize a variety of spatial analysis tools to determine the nature and extent of a flooding event for a specific geographic area. The accuracy of these data, which provide the input into the model, influences the validity of the modeling outputs.

Critical Thinking: Spatial analysis is dependent on the identification of accurate timely data and appropriate tools for manipulating the data to ultimately show where flood waters will go over time and the depth of the water in a spatial context. The methodology that we establish must include the identification and selection of an appropriate data set that can support the results of our analysis.

Riverine flood modeling addresses the question of just how deep the water will be at a given time and location. The map shown in Figure 4.1 provides an illustration of the use of spatial analysis to show the anticipated depth of water for a 100-year flooding event in a drainage area. The Hazards United States Multi Hazard Flood (HAZUS-MH) Flood model developed by Federal Emergency Management Agency (FEMA) provides the means of utilizing many types of data to characterize riverine flooding events for a specific drainage area.

1. Hydrologic data determines just how much water may be in the water feature for a 100-year event, for example, for a water feature and drainage area. Hydrology is the science that deals with the properties, distribution, discharge, and circulation of water on the surface of the land, in the soil and

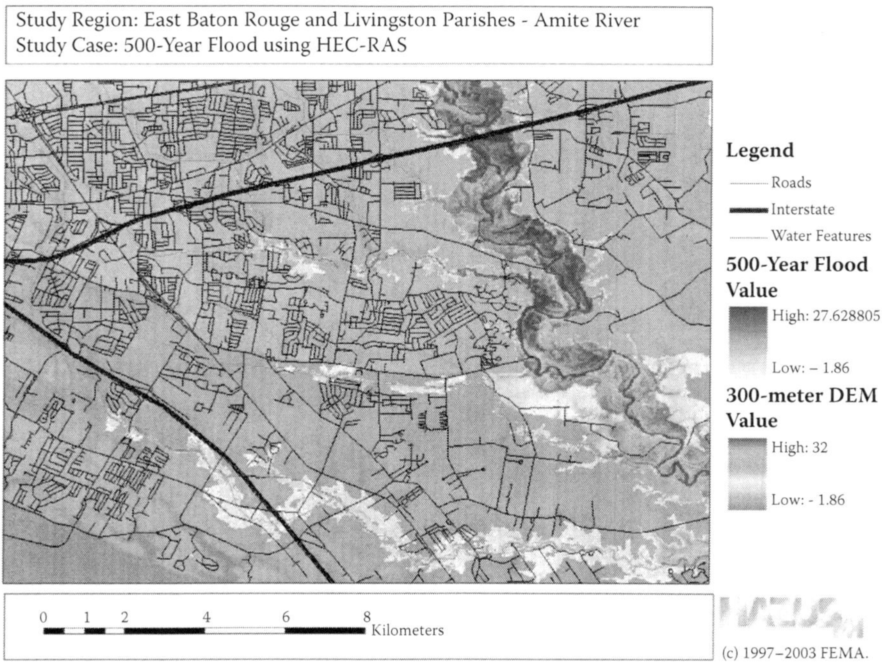

Figure 4.1 (See color insert following page 142.) Riverine flood modeling results within HAZUS-MH Flood.

underlying rocks, and in the atmosphere. It also refers to the flow and behavior of rivers and streams.

2. Hydraulic analysis determines flood elevations for a specific flooding event at a location on a water feature. Hydraulic data thus reflects anticipated areas to be flooded and the depth of flooding. These calculations are determined by comparing the "modeled" flood elevations along a water feature with land contours (Digital Elevation Model [DEM] land elevations). A hydraulic model such as HEC-RAS is used by FEMA and the U.S. Army Corps of Engineers to prepare community flood maps for the National Flood Insurance Program (NFIP). How will the water move and flow in the drainage area? What will be the depth of the water?
3. High-resolution land contour data is obtained from remote sensing tools such as light detection and ranging (LIDAR) flow by either fixed-wing aircraft or helicopters.
4. Spatial modeling tools such as HEC-RAS calculate the depth of water along the water feature. GIS tools depth would need to be calculated on the banks of the water feature and in the deepest areas of the water bed.
5. Location of bridges or culverts that might limit or constrict the flow of the water.

Hurricanes Katrina and Rita provided a unique opportunity for researchers to have access to a large collection of hazard-related data. The Katrina and Rita Geospatial Data Clearinghouse houses numerous data sets that can be used to gain a better understanding of the nature of these two hurricanes and their environmental, social, economic, and physical impacts. Included in the clearinghouse are extensive collections of digital remote sensing data including:

1. High-resolution commercial and government remote-sensing photos of impacted areas a few days following the landfall of each storm (resolution 1:6 inches).
2. Satellite radar data from Radarsat allow users to identify areas that experienced flooding or environmental contamination from oil storage or platform spills.
3. Aircraft LIDAR high-resolution landscape elevation data in coastal areas that allow for examination of land loss issues associated with coastal storms.
4. Moderate resolution imaging spectroradiometer (MODIS) and LANDSAT satellite data that allow for an assessment of land use changes in coastal areas.

For further information on data available from the Katrina and Rita Geospatial Clearinghouse see http://www.katrina.lsu.edu.

Figure 1.3 Louisiana's Comprehensive Master Plan for a Sustainable Coast (http://www.lacpra.org/index.cfm?md=pagebuilder&tmp=home&nid=24&pnid=0&pid=28&fmid=0&catid=0&elid=0).

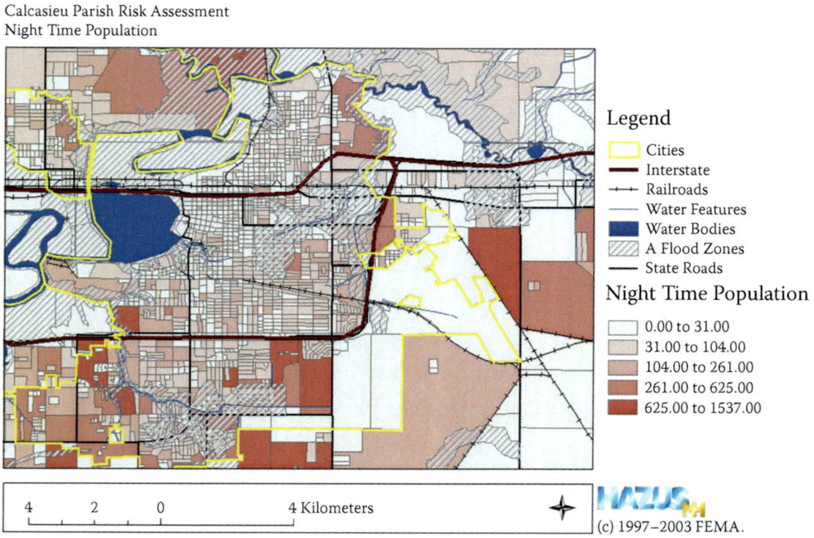

Figure 7.4 Nighttime population with flood zones.

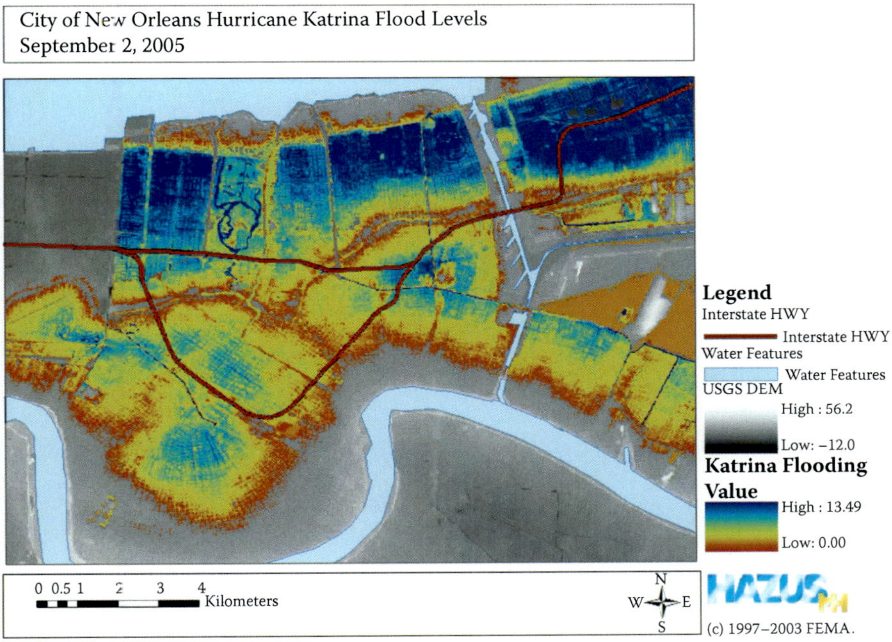

Figure 2.2 City of New Orleans, LA, elevation map.

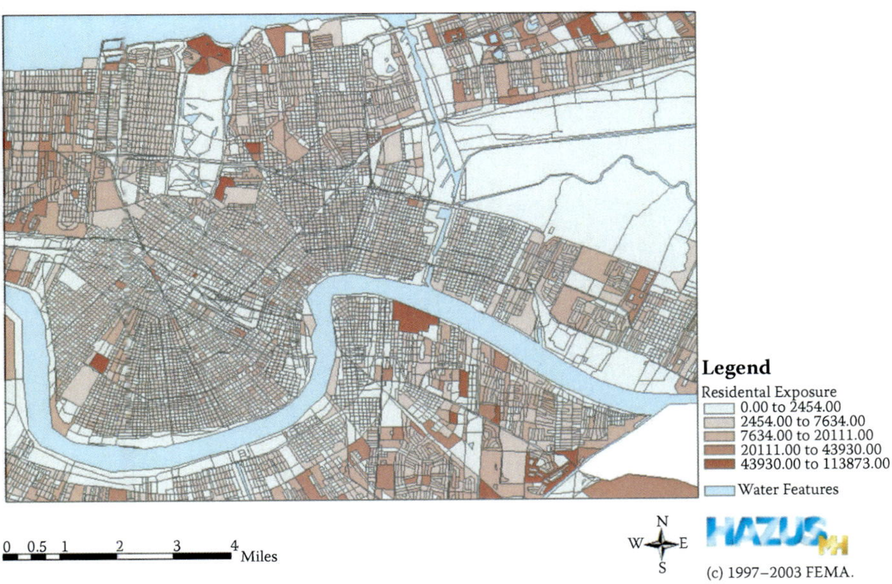

Figure 2.3 A choropleth map of New Orleans, LA, showing residential structure values.

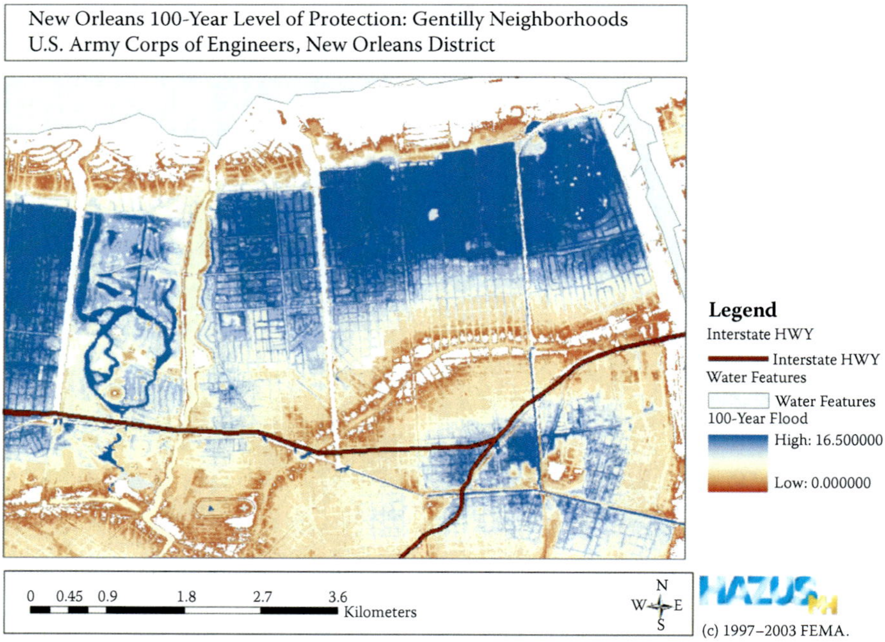

Figure 3.2 Flood map of New Orleans—Gentilly neighborhood (http://www.mvn.usace.army.mil/hps/100maps.htm).

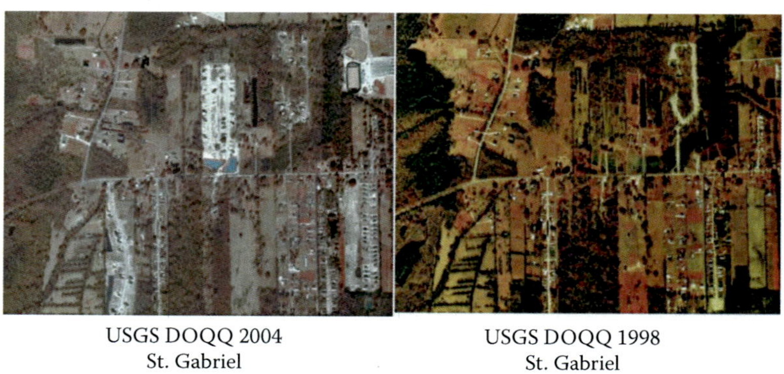

Figure 3.3 Development in a rapidly growing community.

Figure 3.5 USGS DEM, 5-meter DEM, and high-resolution image.

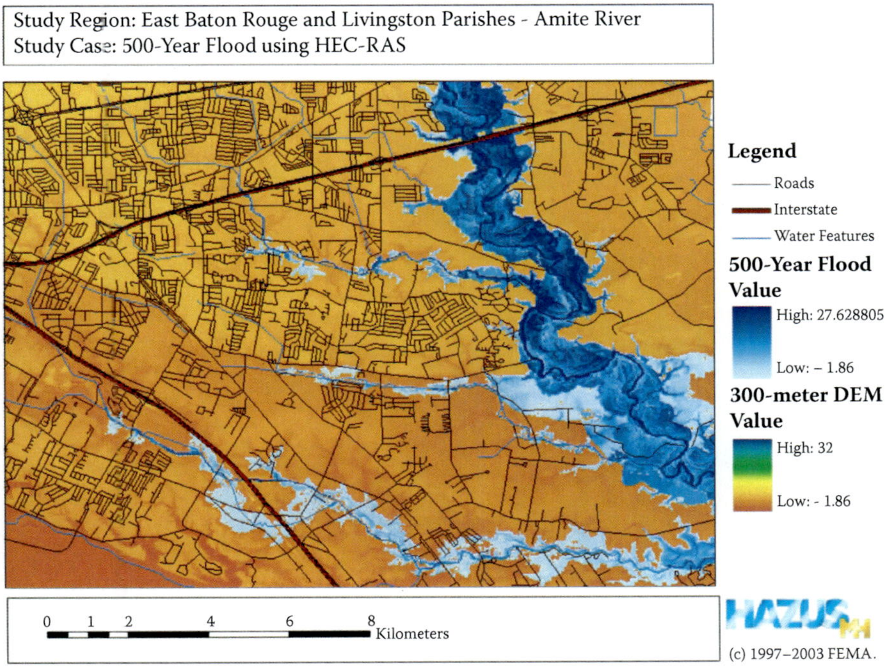

Figure 4.1 Riverine flood modeling results within HAZUS-MH Flood.

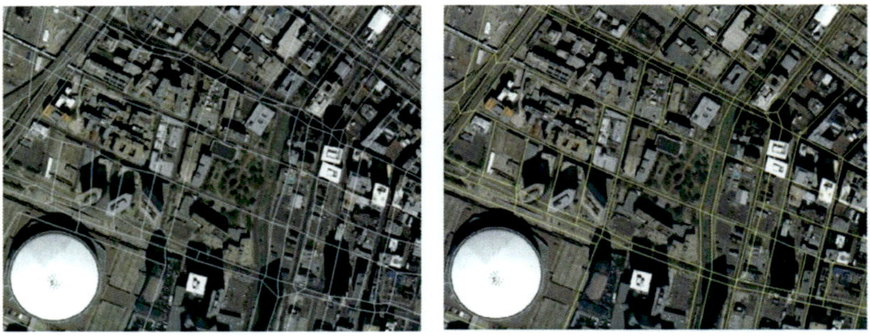

New Orleans High Resolution Image with Census Roads

New Orleans High Resolution Image with Edited Roads

Figure 4.2 Comparison of Census Bureau road files and edited files.

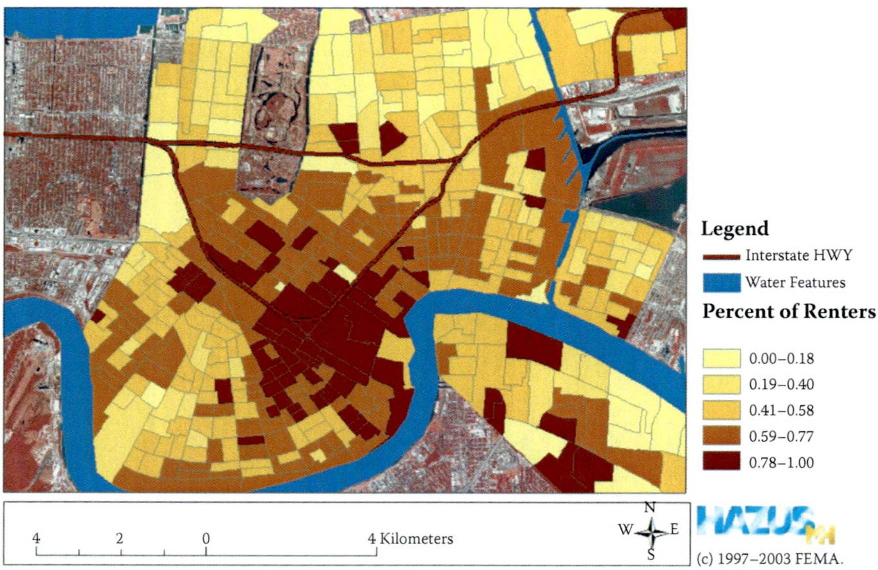

Figure 4.3 Percent of renters for the City of New Orleans at the census-block-group level.

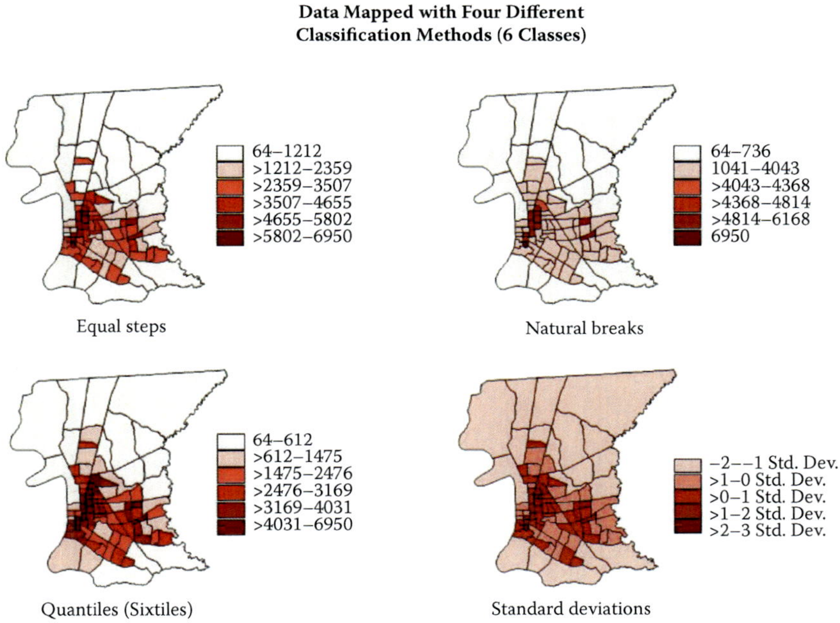

Figure 4.6 Visualization of data using different classification methods.

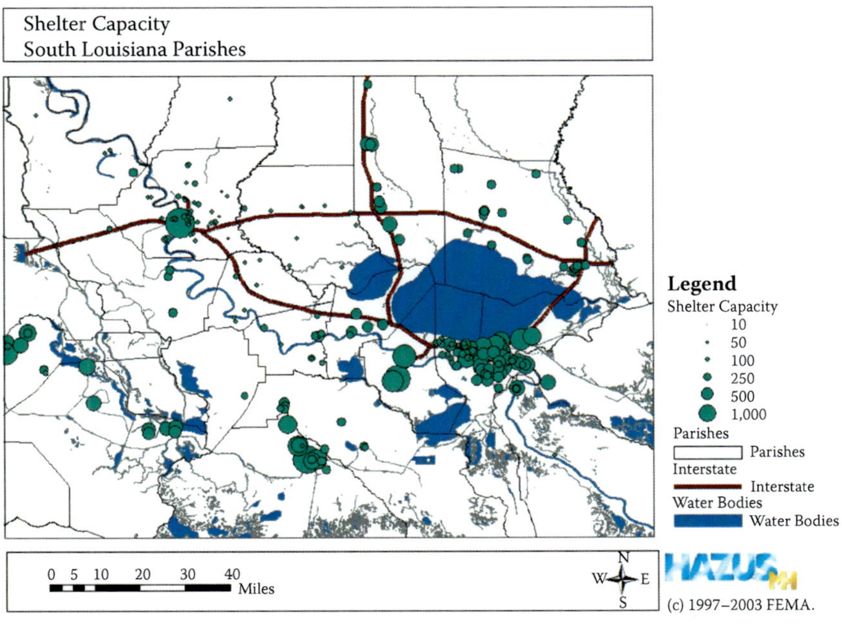

Figure 4.7 Use of proportional symbols in mapping data.

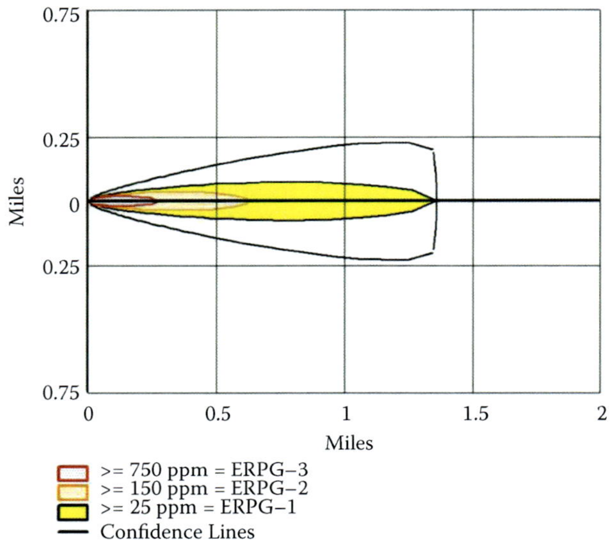

Figure 5.3 Hazard risk zones representing alternative exposure limits.

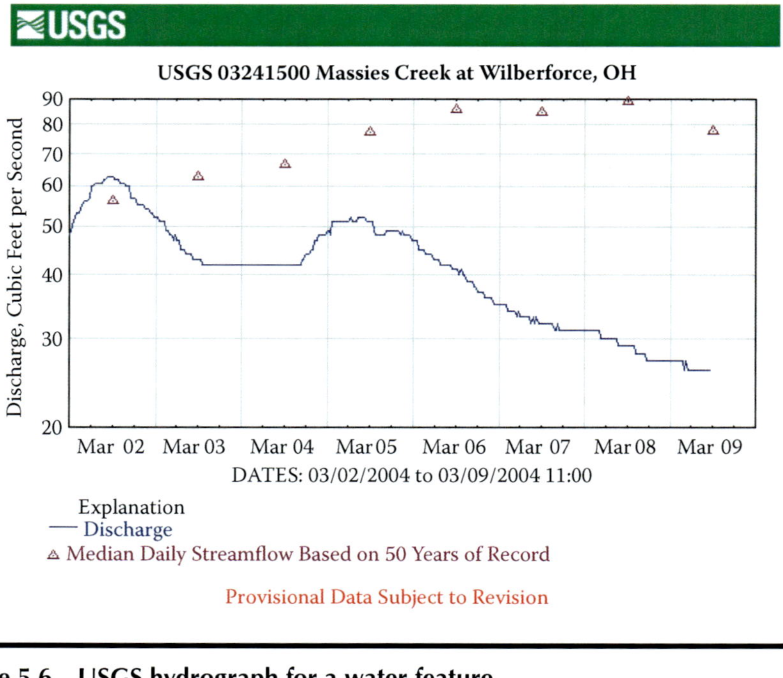

Figure 5.6 USGS hydrograph for a water feature.

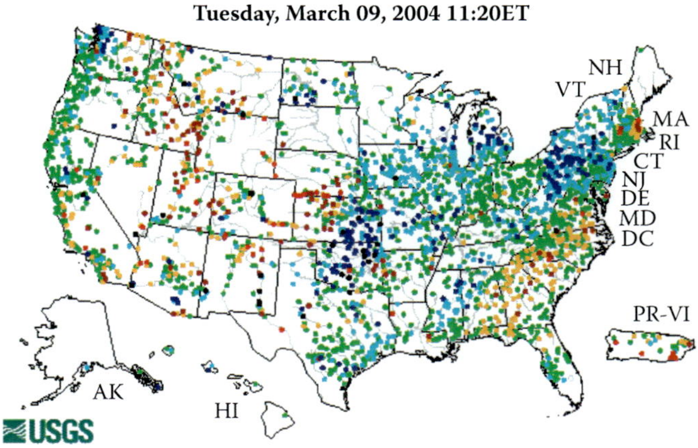

Figure 5.7 USGS river gauges in the United States (go to http://water.usgs.gov/waterwatch to review active state stations).

Figure 6.3 Environmental capital: healthy forest, clean water, and soils that support fish and wildlife.

Spatial Data Quality

We should acknowledge that any data set will not be 100 percent accurate. Errors and uncertainty are inherent in any data set or information system (Openshaw and Clarke 1996) and should be acknowledged as part of the hazards analysis process.

Critical Thinking: To what degree does our data set accurately represent our environment (social, economic, ecological, and built)? Understanding the limitations of the data set is critical in formulating a sound methodology for our hazards analysis. What special problems are present in data sets? How does the availability of data influence our methodology that we use in our hazards analysis?

Many users of hazards analysis inherently trust computer outputs, especially in a complex environmental hazards analysis. We should acknowledge that the computer model is just a tool that includes assumptions about the environment and the relationships between its variables. We should be very clear as to the limitations of the data inputs and the assumptions that the model makes in simulating a complex environmental hazard.

Those that use the outputs from a hazards analysis that are from nonspatial disciplines need to appreciate the uncertainly that is inherent in spatial data sets and the consequences of using these data sets in our analysis. There are clear limitations in any data set used in a hazards analysis; clearly expressing these limitations is critical for an appropriate application of the hazards analysis outputs in decision making. Goodchild (1993) stresses that GIS layers have inherent errors that may be obvious to GIS specialists but not understood or appreciated by those from other disciplines. The key is that one should not ignore inherent errors that are just part of a geospatial dataset. Errors may occur in either the source of the data or in the processing steps of the GIS.

Hazards analysis combines the use of GIS (including spatial analysis), environmental modeling, and remote-sensing data sets. The linkages and integration between them are evolving, and weaknesses exist. Clarifying how these tools have been integrated must be explained in our methodology for a hazards analysis.

Data is collected within a specific context. Metadata files associated with a data set describe the process of the data collection, the purpose of the data set, time lines for data collection, processing, assessment, and distribution. Understanding why and how the data was collected must be part of the methodology for our hazards analysis. Any conflicts that are identified with the scope and purpose of the data set and our use of the data must be explained.

We stress that our spatial analysis approach or methodology must include an examination of our metadata files, which document, who established the data set, when, the intended use, and date of outputs, but which rarely address the accuracy of the dataset. The metadata will provide us the information to explain why our selected data is suitable for what we hope to accomplish in our hazards analysis.

This may be because it is just too costly to assess the spatial error in the data or because of the complexity of completing such an assessment.

Key terms that help us understand spatial data quality include error, accuracy, precision, and reliability. Error is any deviation of an observation and computation from what exists or what is perceived as truth (Brimicombe 2003). Accuracy is the degree of fit between our observation or computation with reality. Precision is the degree of consistency between our observations and what exists in the natural, social, or built environments. Reliability involves our confidence in the fit between our data and our intended application of the data in the hazards analysis process. For our purposes, the persons responsible for establishing a methodology for a hazards analysis has the responsibility to articulate what data sets are being used in our analysis and why we believe they are appropriate for our use. Our judgment as to the reliability of the data sets is critical in ensuring that users of the hazards analysis have confidence that their decisions are sound and can be supported by our methodology. Uncertainty is an inherent element of the hazards analysis process from the ways that we obtain data sets, use them, store and manipulate them, and present the results of our analysis as information in support of organizational decisions. The outputs from our hazards analysis are thus dependent on data quality and model quality (including any spatial, statistical, or GIS tools that we utilize) (Burrough et al. 1996).

A few illustrations can help demonstrate the importance of understanding the purpose of and intended use of a dataset, who collected it, when it was collected, and how and when it was disseminated. Many community and organizational hazards analysis utilize Census Bureau road, water feature, community boundary, and point files. The Census Bureau obtained these critical community files from the U.S. Geological Survey and added critical data to the lines (roads, rail lines, and water features), points (community features, such as schools, churches, or public buildings), and polygons (lakes and political boundaries). The Census Bureau has over many years made changes to these files to more accurately reflect what exists throughout the United States. A process of involving community partners in updating these files has resulted in very accurate data sets for some communities.

The Census Bureau has obtained from many local communities' updated road files, school, medical facilities and church locations, and political boundaries. The names and addresses of schools, medical facilities, and churches may have changed over the past fifty years and new road features added in a community. Many local governmental emergency communication districts have taken the Census Bureau road files and aligned them over very high-resolution digital images of their community. For many communities, high-resolution images of a half-foot resolution provide a basis for ensuring that a road feature or a school location is highly accurate. Prior to these corrections being made, the Census Bureau map files so often used in a hazards analysis have extensive errors in the name of a specific feature and its location. It is not uncommon that a road or other feature may be off by as much as 100 feet when observed on a high-resolution image of a community.

Unfortunately, easy-to-use GIS programs were not available to local communities when the Census Bureau created the map files that have been used as part of the Centennial Census. Errors thus could be present either in geographic representation of the object or because of errors in the attributes reflected in the data (i.e., the road name or feature name is incorrect).

We should not avoid using Census Bureau map files in our hazards analysis, but insist that the metadata be reviewed and that any errors inherent in the road or water feature files be fully understood and explained in our methodology used in the hazards analysis.

Figure 4.2 provides a comparison between common community road files obtained and edited road files over a high-resolution image. The image was taken after Hurricane Katrina in January 2006. These road files had been edited by the New Orleans Regional Planning Commission GIS unit for the City of New Orleans Planning Department years before Hurricane Katrina struck south Louisiana. These edited street files have been a long-standing asset to local and regional hazards analysis efforts in the public, private, and nonprofit sectors in the New Orleans area. The unedited Census road files on the left are the type of road files that are available from many sources that are commonly used by local jurisdictions as part of their base map. The edited files have been corrected using high-resolution images such as the ones above. They provide a highly accurate basis for spatial analysis. High-resolution photos were not available when the United States Geological Survey (USGS) created the road and street files that were later used by the Census Bureau as a guide for census workers to navigate local communities. Many communities have edited the Census road and street files so that they more accurately reflect the local landscape when imposed over high-resolution images. Users of data such as Census road files must appreciate the errors that exist in the files and if they are an appropriate basis for analysis of hazards at the community level.

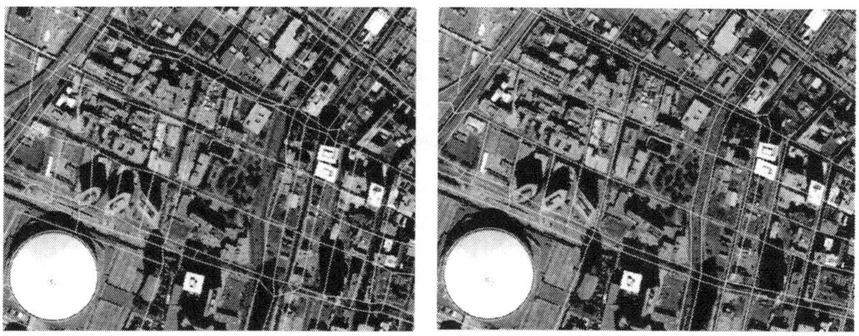

New Orleans High Resolution Image with Census Roads

New Orleans High Resolution Image with Edited Roads

Figure 4.2 (See color insert following page 142.) Comparison of Census Bureau road files and edited files. High resolution image provided by NOAA (2005).

Critical Thinking: If a local community was utilizing unedited Census road and street files along with Census population data at the block or block group resolution, could the data sets be used without potential errors distorting the results of the analysis?

Many errors that are inherent in data sets used in a hazards analysis occur because of changes over time. Changes in water features, land use, or landscapes may occur naturally or because of human interventions. We must examine any data set that is part of our hazards analysis to understand if changes have occurred and that these are noted in our methodology.

It should be noted here, as we discuss data quality, that modelers assess the quality of their outputs by comparing the results of simulated disasters with actual events. The National Weather Service (NWS) has customarily compared model results from hurricanes with weather data from sensors in coastal environments. These comparative studies provide the basis for adapting hurricane models and improve their predictive capacity for future storms. Post-Rita and Katrina storm surge measurements in 2005 provided modelers running the Advanced Circulation Model for Coastal Ocean Hydrodynamics (ADCIRC) hurricane model with invaluable measurements to compare simulated surge heights at specific locations with actual storm surge heights in coastal Mississippi and Louisiana.

Types of Spatial Analysis

Queries

How many people, commercial businesses, or residential homes might be impacted by a flood or storm surge? How many roads or bridges are in the area with the deepest flooding? How many structures are in the high-wind zone of a hurricane? How many renters or homeowners may be displaced by a flooding event? What is the average income of population of a community directly impacted by a hurricane? How many employees are affected by businesses in a flood zone?

Spatial analysis can address the question of access to major transportation routes by renters, households below the poverty level, households with no automobiles and access to public transportation routes, or households and shelters with handicapped individuals over the age of 65. Spatial analysis provides a means of comparing renters and homeowners and access to evacuation routes, evacuation access points, or shelters. Figure 4.3 shows the percent of renters by census-block-group level in New Orleans. The analysis could help determine if renters might be more vulnerable than homeowners if an evacuation was ordered. Further, the analysis would be able to show which block group areas for either renters or homeowners are at higher risk by living further from an evacuation route, pick-up point, or shelter. With this information, emergency management staff could target

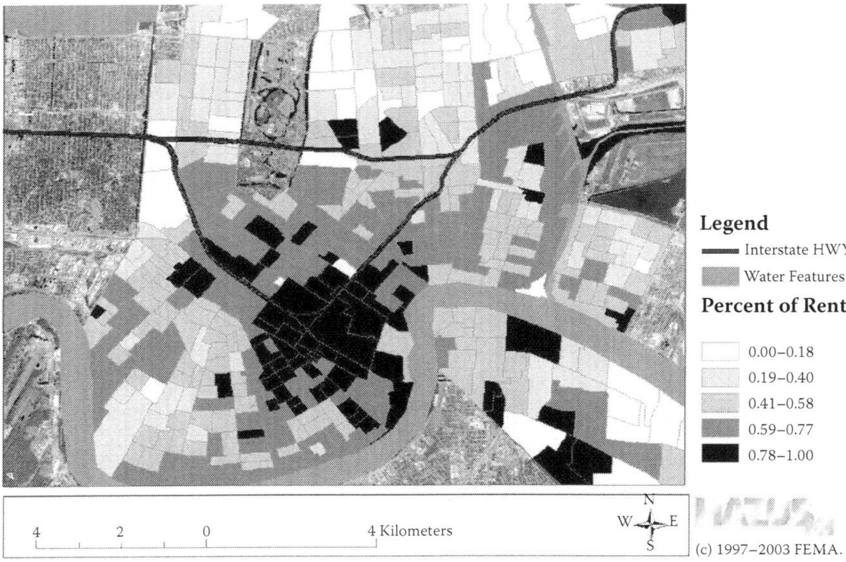

Figure 4.3 (See color insert following page 142.) Percent of renters for the City of New Orleans at the census-block-group level. Background image provided by the City of New Orleans.

specific areas of the community for a contingency plan to ensure that all residents would be safe in an emergency.

Measurements

Hurricane Katrina flooded many communities in the greater New Orleans area. What was the area flooded in the City of New Orleans? How did this change as rescue efforts progressed and pumps were used to remove the water? What is the average residential parcel or lot size in flood areas of the city? Using land-use classification data for the City of New Orleans, how much commercial or industrial property was flooded? How much public property for parks and open space was flooded? How much of the city's poor neighborhoods were flooded as compared to more wealthy areas?

With the flood depth grid shown in Figure 4.4 for the City of New Orleans during Hurricane Katrina, one could use spatial analysis to determine if a higher percentage of households in flooded areas had incomes below the poverty level, had no access to an automobile, were handicapped, were renters, or had a single head of household with children below the age of 18. Pedro (2006) examined these questions in her Master of Science thesis, using the flood depth levels from a hurricane

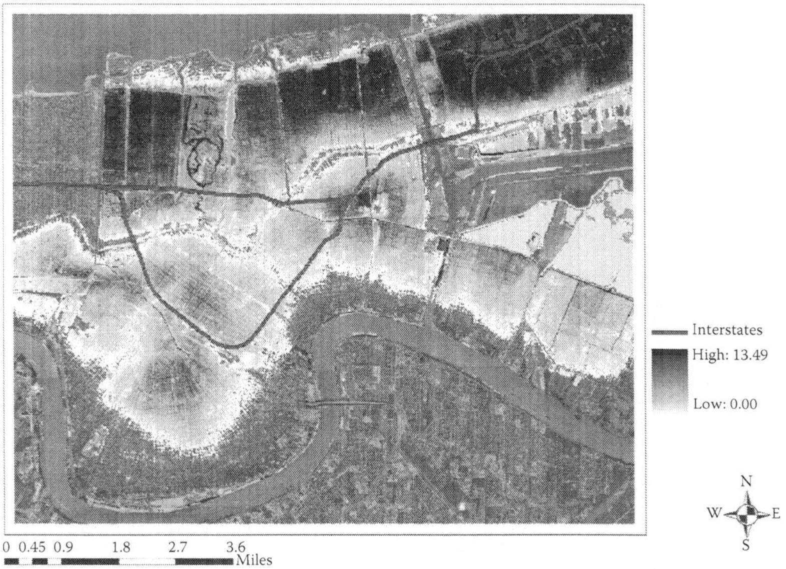

Figure 4.4 City of New Orleans flooding following Hurricane Katrina (NOAA figures).

simulation for 2005, and determined that there were no differences between households on these characteristics when comparing flooded and nonflooded areas of the City of New Orleans. She also addressed the hypothesis that the simulated flooding did not have a disproportionate impact on the percentage of households in census block groups who were African American, below the poverty level, renters, or who did not have an automobile. With this analysis, she was able to pinpoint areas of the City of New Orleans in which the depth of flooding might be very high and a greater percentage of residents would not have access to an automobile, were below the poverty level, and had a single head of household with children under the age of 18.

Spatial analysis provides us with a set of tools with which we can explore questions about potential vulnerability to and damage from hazards. It can provide information that can be used to take precautionary measures or to further explore if some neighborhoods were more vulnerable than others and if assistance with evacuations or sheltering was needed.

Transformations

These analysis tools allow the user to transform GIS data sets to reveal relationships and dynamics of the physical environment. Examples include buffering a point, line, or area to highlight potential change. If a new school were to be built in a specific location, what is the population in a two-mile area? If a commercial area were to be flooded, what other enterprises in a three-mile area could handle the additional business? If a rail line was damaged as a result of an earthquake, how many industrial enterprises within a ten-mile area could be impacted?

Buffering

Buffering was used by Pine et al. (2002) to determine if African Americans were in closer proximity to thirteen large chemical processing sites in Iberville Parish in Louisiana. The question centered on whether African American residents were not closer to chemical processing operations than non-African Americans. The buffer zones were determined using dispersion modeling programs from the thirteen sites, and the number of African Americans was calculated for each risk zone from Census 2000 data. The buffering spatial analysis tool was helpful in examining claims of disparate impact of chemical releases for African Americans in a community. The study showed that African Americans did have a greater chance of living closer to one of the thirteen chemical processing operations than non-African American residents (Figure 4.5).

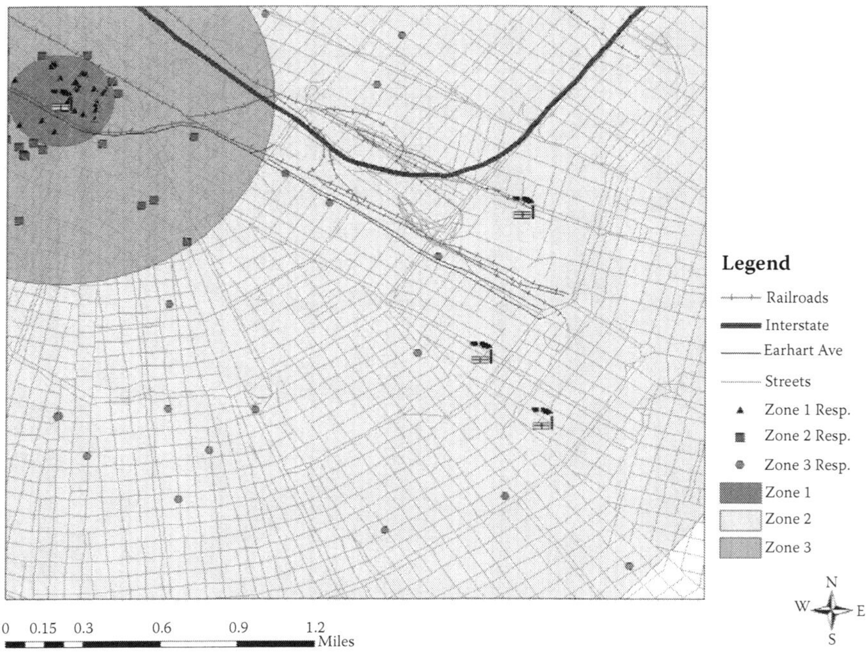

Figure 4.5 Race and distance.

Spatial interpolation is used to help estimate potential flooding along a water feature where hydrologic modeling programs determine the depth and extent of flooding at various locations along a stream, bayou, or river. Riverine flood models utilize precise elevation measurements along water features as part of the flooding program. Spatial interpolation is used to estimate the depth and extent of flooding between survey points. Many hydrological flooding efforts include field survey cross sections along water features as a basis for determining flood depths. The depth of flooding between the cross sections is interpolated spatially. The model thus produces a smooth flood zone and depth of flooding for the area impacted using this spatial analysis tool.

Descriptive Summaries

Data sets that reflect unique elements of a disaster provide opportunities for understanding potential relationships between a disaster and associated human characteristics. Following Hurricane Katrina, 911 emergency calls at point locations were examined to see if the calls were clustered in some way to suggest the basis of the emergency call. The issue centered on whether there was a relationship between clusters of 911 emergency calls and water depth and selected social characteristics. Spatial analysis was used to identify hot spots areas where there was a high number of emergency calls for assistance. Further analysis examined the depth of the water at these hot spots and social characteristics of the census tract.

Optimization Techniques

Spatial analysis is also used in site selection and transportation routing to help locate the ideal setting for an emergency shelter, medical clinic, or police substation, or the shortest evacuation route of multilane roads and highways. Evacuation routes that are scenario specific can be developed to aid community planners in evacuating large populations from a metropolitan area. State departments of transportation have used these tools to mark major evacuation routes as aids to move citizens from vulnerable areas due to hurricanes, earthquakes, or riverine flooding.

Hypothesis Testing

This type of spatial analysis was used in New Orleans to anticipate the rate of return to specific neighborhoods following the flooding from Hurricane Katrina. A statistically valid sample of household surveys was conducted to determine the family's capacity and willingness to return to the city after their residence had been flooded. The address of the respondent was obtained in the survey along with the resident's perception of the level of damage to their home and neighborhood. Independent surveys of residential structures conducted by FEMA, the City of New Orleans, and the Louisiana Road Home Program provided an independent perspective on individual property damage. Household sentiment to return was then compared to their perception of the level of damage to their home and the independent property damage assessment. Spatial analysis was utilized to use the results of the surveys to infer if other residents would return in a given time period. This type of analysis provided a basis for testing a set of hypotheses relating to the willingness to return to a specific structure or neighborhood.

Spatial analysis was used to examine a hypothesis concerning social vulnerability and flood depth following Hurricane Katrina. The question centered on an association between risk zones measured by depth of flood waters and social vulnerability. This study, which is under review for publication, examined the relationship between risk (i.e., flood depth) and social population characteristics, including race, income, disability, home ownership, single family member and head of household, and household access to an automobile. The analysis revealed that African Americans had the strongest association with deeper flood waters when compared with other populations characteristic for the Orleans Parish. The results of the analysis is being used by City of New Orleans emergency management officials to identify neighborhoods that are highly vulnerable to future flooding, and family evacuation plans are prepared in advance of coastal storms.

Since two different data sets were used in the study, one had to be converted to a common type and scale. Flood depth values from a grid file were selected and then averaged for each census block group. Since land elevations in New Orleans varied only slightly within a block group, this conversion produced good flood estimates for each block group. It is critical that the methodology used in a hazards analysis fully explore any potential problems with geospatial data that is utilized and explain the source of any data and how it may have been adapted for the spatial analysis.

Spatial Data Visualization

A critical part of the hazards analysis process is displaying the results on our analysis. Hazards are geospatially oriented, and thus being able to show the results of our analysis is a key element in supporting individual, organizational, and community decision making. Whether the results of our analysis comes from a modeling program such as HAZUS-MH, earthquake, flood, coastal hazards, or Areal Locations of Hazardous Atmospheres (ALOHA), displaying geospatial data is a critical means of conveying information to users of our final hazards analysis or as we work with the data using spatial analysis techniques. Spatial data is meant to be viewed as maps, and a GIS allows us to interactively change these maps to help us reveal information from the landscape in a variety of ways. We can add different spatial layers, such as where people live, transportation routes, quarantine areas, key facilities, or infrastructure, and display them over high-resolution images of a community.

Visualization of the results of a hazards analysis and the use of spatial analysis and mapping tools can help us to:

1. Identify patterns within complex data sets or multiple data sets of related data.
2. Make sense of large data sets.
3. Appreciate that local geospatial features change over time.
4. Appreciate that geospatial features may be similar, or interact more frequently, within smaller geographic scales.
5. Provide a means of conveying complex information without oversimplification of the data.
6. Give an Emergency Operations Center (EOC) at a local, state, or regional level critical information on the nature of hazards and their potential impacts.

Both Figure 4.3 and Figure 4.4 offer illustrations of how we visualize hazard vulnerability in a simulation or exercise or an actual disaster response. In an examination of the map of the City of New Orleans in Figure 4.3, one sees that there is great variation in the percentage of the population that are renters. Two patterns may be seen to suggest that the higher percentage of census block groups are in the central and midcity neighborhoods, while those areas on the urban fringe of New Orleans have the lowest percentage of renters. A test of this theory and the association between income and percent of renters can be determined quantitatively using spatial analysis. Maps as in Figure 4.4 provide us with a broad view of the community and a basis for forming our hypothesis, which may be tested using spatial analysis.

Additional hypotheses could be identified, using broad views of the community, raising questions as to the relationship between ground elevation and household income, the relationship between major transportation routes and rental housing, or the association between housing values and community recreation areas (parks). A look at Figure 4.3 shows that City Park is located in the top of the image and that limited rental units are available near the community recreation area. A spatial analysis could address the hypothesis to determine if there is a clear association between large community recreation areas and block groups with low percentages of households of renters.

Figure 4.4 was used extensively by emergency responders at the local, state, and federal level following the flooding from Hurricane Katrina. As water depth changed as the pumping process was underway, one could provide information to emergency responders on major transportation routes and flood depth. Spatial analysis was used to determine the best rescue routes throughout New Orleans.

For any hazard, there is both a spatial and temporal dimension. The spatial dimension has various scales (local to international) depending on what the hazard is. The temporal dimension also has multiple scales (minutes to months).

Mapping data related to a hazard utilizes many different data sets and types of mapping data including photographs, USGS quad sheets showing graphic representation of areas, and vector representations of roads (lines), buildings (points), and county or city boundaries (polygons). What are the best ways to show hazards from wind, flooding, storm surge, earthquake, landslide, drought, or other disaster?

Every map is a graphic representation or a model of reality or milieu:

1. A map may represent economic or cultural features, such as neighborhoods or citizen sentiment, settlement patterns, and political–administrative boundaries.
2. A map represents physical features, such as elevation, water features, and land cover.

A map can also display mental abstractions that are not physically present on the geographical landscape. For example, we can map people's attitudes, quality of life, or citizen sentiment for or against gun control.

Choropleth Maps

A choropleth map is defined by the International Cartographic Association (ICA) as "a method of cartographic representation, which employs distinctive color or shading to areas other than the feature boundaries. These are usually statistical or administrative areas." Making a choropleth map starts with the collection of data by a specific geographic area. An areal symbolization scheme is then devised for these values, and the symbols are applied to those areas on the map whose data fall into the symbol classes. The selection of symbol classes is based on a classification method.

It is important that choropleth maps show relative data in contrast to absolute data. Relative data includes densities (e.g., population density—people per square mile), percentages (e.g., percent of people 65 years and older), and rates (e.g., number of homicides per 100,000 people). The following three decisions have to be made when compiling any choropleth map:

1. Number of classes. A trade-off exists between too many and too few classes. Too many classes make the choropleth map too complex and difficult to perceive and to understand by the map reader. Too few classes results in too much information loss.
2. Type of classification method. This includes equal steps, which could be intervals such as 0–10, >10–20, >20–30, etc. A second one involves natural breaks in the data, which are reflected in the sorted data. Quantiles could be used as a classification method using equal number of observations in each class. Finally, standard deviation can be used, which includes the average deviation of the data values from the mean (average) of the data set. This approach measures the variability in the data and is a good relative measurement tool.
3. Color or areal symbolization scheme. This can be used to simply show the different areas (zip codes, incorporated areas, or districts) in various colors.

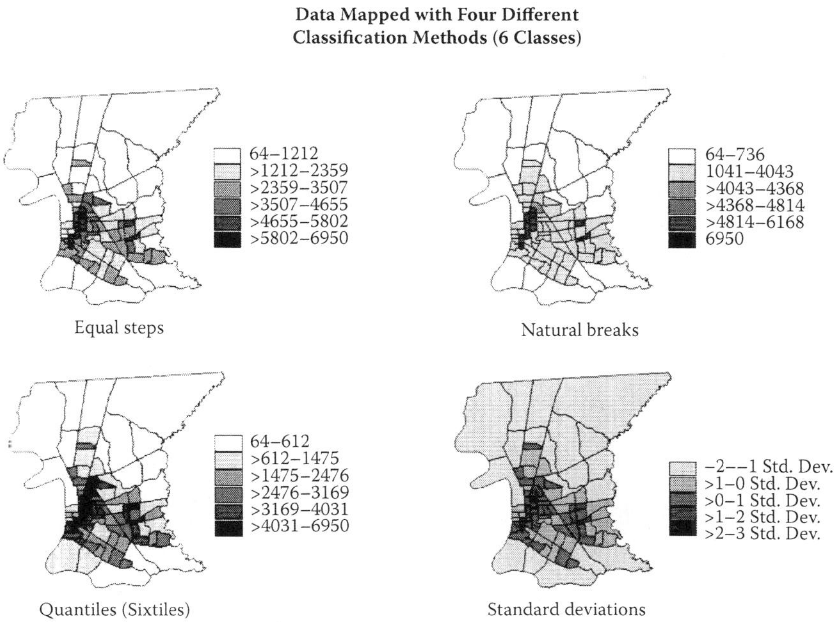

Figure 4.6 (See color insert following page 142.) Visualization of data using different classification methods.

Critical Thinking: Figure 4.6 provides an example of how we can display the same information using different classification methods. The maps show that the manner in which we view the data will influence the conclusions that we draw. Which of the four maps provide the best spatial and temporal perspectives?

An example of a proportional symbol map is the common dot map. Common dot mapping involves the selection of an appropriate point symbol to represent each discrete element of a geographically distributed phenomenon. The symbol form does not change, but its number changes from place to place in proportion to the number of objects being represented. Design decisions involve the placement of dots and the selection of dot value and dot size. Figure 4.7 shows data in a jurisdiction that could have resulted from an analysis of spatial data. The dot symbols reflect data for a specific geographic boundary.

A proportional symbol map uses a form (circle, square, or triangle) and varies its size from place to place, in proportion to the quantities it represents. Map readers can form a picture of the quantitative distribution by examining the pattern of differently sized symbols. Proportional point symbol mapping is selected when data occur at points or when data is aggregated at points representing areas as illustrated in Figure 4.7.

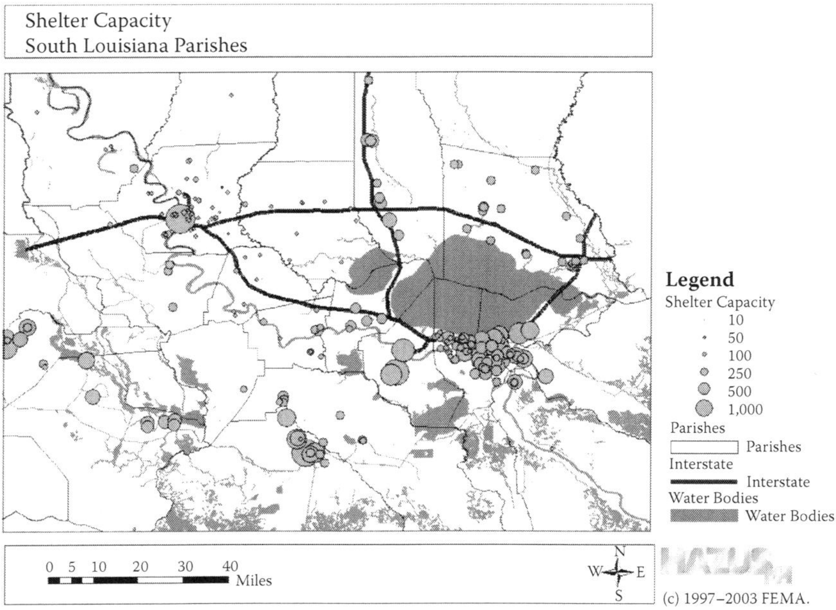

Figure 4.7 (See color insert following page 142.) Use of proportional symbols in mapping data.

Conclusions

Brimicombe (2003) stresses that suitability of a dataset for our use in a hazards analysis centers on its "fitness for use." Rather than focus on errors, he encourages us to view our use of data in a wider examination of uncertainty. We should examine the quality of the data, but explain that the dataset is an appropriate application for our use. We thus examine the quality of the dataset and how we plan on using the data in our methodology or our approach to spatial analysis and our overall hazards analysis. In the end we want to ensure that our data fits our methodology and is an appropriate use of the dataset in our analysis. A fitness-for-use test thus includes an evaluation of the data quality and an explanation of any limitations of the data as it is used in our hazard models or spatial analysis techniques. The key is that this is a managed process that accounts for limitations in our data. Our methodology in completing the hazards analysis should clearly explain the steps that we are taking in the spatial analysis, including the source of our data and our analysis of this data.

We stress that our methodology must include an examination of our metadata files, which document who established the data set, when, the intended use, and date of outputs, but rarely address the accuracy of the dataset. The metadata will provide us the information to explain why our selected data is suitable for what we hope to accomplish in our hazards analysis. This may be because it is just too costly to assess the spatial error in the data or because of the complexity of completing such an assessment.

Discussion Questions

Many efforts to examine environmental hazards require the use of digital elevation map data (DEM). The USGS has published for many years DEM data for the United States in different scales (1:30 meters or 1:20 meters). More recently, many state and federal agencies have created higher-resolution DEM data using LIDAR remote-sensing technology (1:5-meter resolution). What difference does using the higher-resolution DEM make for an environmental hazards analysis?

Given that any dataset will not be 100% accurate, why is it so important that errors be examined and explained in a hazards analysis?

What do the terms error, accuracy, precision, and reliability mean, and why are they so critical to the use of data in a hazards analysis?

How might the results of a spatial analysis be visualized to communicate information that can help us to understand the outputs of a hazards analysis?

Applications

Metadata provides information about a geospatial data set to guide the user in determining how best to utilize the data in a hazards analysis or other application. Metadata files are provided for many data sets as illustrated by the Atlas Internet site at Louisiana State University (http://atlas.lsu.edu). Go to this site, select the LIDAR data set, and download for any geographic area of the state one metadata file. Read the file and determine how the data was obtained, who collected the data, when was it collected and made available to the public, what the resolution of the data is, and what the intended purpose of the data is. The file notes that no data quality tests were performed on this data. What might that mean to the user of the data?

Ask students to use any disaster data and visualize them with a proportional point symbol map.

Map a disaster using a common dot map approach for the same study area, but using different dot sizes and dot values, and observe the differences between the different dot maps. Use any mapping or GIS software package to perform this exercise.

Web Sites

FedStats. Federal Statistics. http://www.fedstats.gov/

Geodata.gov. GOS—Geospatial One Stop. http://gos2.geodata.gov/wps/portal/gos
U.S. Census Bureau. TIGER/Line shapefiles. http://www.census.gov/geo/www/tiger/tgrshp.html

References

Brimicombe, A. (2003). *GIS, Environmental Modeling and Engineering.* Taylor and Francis. London.

Burrough, P. A. and A. U. Frank, eds. (1996). *Geographical Objects with Indeterminate Boundaries.* Taylor and Francis, London.

Fischer, M. M., H. J. Scholten, and D. Unwin, eds. (1996). *Spatial Analytical Perspectives on GIS.* Taylor and Francis, London.

Gilbert, E. W. (1958). Pioneer maps of health and disease in England. *Geographical Journal,* 124:172–183.

Goodchild, M. F. (1993). Data models and data quality: problems and prospects. In *Environmental Modeling with GIS,* eds. M. F. Goodchild, B. O. Parks, and L. T. Steyaert. Oxford University Press, New York.

Longley, P. A., M. F. Goodchild, D. J. Maguire, and D. W. Rhind (2005). *Geographic Information Systems and Science,* 2nd Edition. John Wiley and Sons, New York.

Openshaw, S. and G. Clarke (1996). Developing spatial analysis functions relevant to GIS environments. In *Spatial Analytical Perspectives on GIS,* eds. M. Fischer, H. J. Scholten, and D. Unwin. Taylor and Francis, London, 21–37.

Pedro, S. (2006). Delineating Hurricane Vulnerable Populations in Orleans Parish, Louisiana. Thesis for an M.S. degree, Louisiana State University.

Pine, J. C., B. D. Marx, and A. Lakshmaṇan (2002). An examination of accidental release scenarios from chemical processing sites: the relation of race to distance. *Social Sciences Quarterly* Special Environmental Justice Issue 83(1):317–331.

Chapter 5

Risk Analysis

Assessing the Risks of Natural Hazards

Objectives

1. Explain the process of risk analysis.
2. Explain what is risk.
3. Compare and contrast quantitative and qualitative approaches to risk analysis.
4. Identify and discuss approaches related to using historical data in determining risk.
5. Explain the concept of uncertainty and how it impacts risk analysis.
6. Discuss the concept of acceptable risk and how we determine it.
7. Explain how we describe the likelihood and consequences of risks.

Key Terms

Risk
Quantitative analyses and qualitative analyses
Acceptable risk
Voluntary risks
Involuntary risks
Flood flow frequency

Flood discharge
Uncertainty
Ambiguity

Introduction

Once we have characterized the nature and extent of hazards by preparing a community profile and a hazards profile for the hazards in the community, we do an analysis of the risks present by identifying vulnerability indicators and the hazards probability. By doing so we explore the likely impact of the hazards on the community. Risk analysis is the determination of likelihood and possible consequences from a disaster. This analysis must be accomplished in a systematic basis so as to provide useful and accurate information for decision makers through risk management or community hazard mitigation initiatives. Our goal is to provide the right information, at the right level of complexity and detail, to decision makers at the right time.

The Process of Risk Analysis

Risk analysis, therefore, takes information from the hazards identification and examines not only the probability of the event but also explores the social-cultural, economic, and environmental adverse impacts from a disaster. Risk analysis goes further to then compare various hazard risks to determine if various risks could have a similar likelihood of occurrence and outcomes. The results of the risk analysis can be used in the problem-solving and decision-making process to adopt strategies to reduce organizational or community vulnerability.

The process of risk analysis examines the nature of the risk from a hazard, when and where it might occur, potential intensity, and potential impact on people and property. The level of risk for a disaster of any scale is expressed as a likelihood of the occurrence or frequency times its consequences. The level of risk may be very limited so that nothing needs to be done to address it. Other hazards may be more likely to occur and have the potential to cause extensive damages. The fact is that some organizations and communities may be willing to live with a specific risk or not willing to expend the resources necessary to reduce the adverse consequences that come with it (Waugh 2000).

For example, a community may be vulnerable to natural hazards because it is located on a coastline, in the mountains, along rivers and lakes, in wooded areas, or in a desert area and, as a result, subject to hurricane winds or storm surge, earthquakes, floods, fire, and other disasters.

Some organizations may be constrained in the location of their business, where they live or locate their operation. As a result, it may be necessary to locate their operation in a floodplain, on a coast, or near a major transportation route. Individual families, organizations, and communities must make conscious choices about what is an "acceptable risk." Hazard reduction policies can be made with an understanding of what choices are possible and the consequences for any option. The decision-making process might include mapping the hazard to determine the spatial distribution of risk, the collection of data on the frequency and intensity of past disasters, judgments concerning specific risk factors (i.e., factors that may significantly increase or decrease the risk of disaster or the threat to life and property), and the vulnerability of the people and property within the risk area. This is all part of the hazards analysis process.

Risk managers consider the likelihood and consequence of all (identified) hazards faced by their jurisdiction, and they rank them according to priority. However, to understand the likelihood component of the risk analysis, you need to have an understanding of probability. Probability is what tells risk managers whether or not they should expect a hazard to affect their community. Jardine and Hrudey suggest that a classical or frequency concept of probability be used to focus on discrete events, which examines all possible outcomes and the numerical relationships among the chances of these outcomes (1997). In the real world, however, such complete information is seldom, if ever, available. Therefore, risk analyses must include subjective information along with detailed historical information.

What Is Risk?

Risk is the product of likelihood or probability of a hazard occurring and the adverse consequences from the event and is viewed by many as simply our exposure to hazards. Simply stated,

$$\text{RISK} = (\text{LIKELIHOOD or PROBABILITY}) \times \text{CONSEQUENCE}$$

This approach is based on the Royal Society Study Group (1993) defining risk as "the probability that a particular adverse event occurs during a stated period of time, or results from a particular challenge" (1992: 2). The Society provides a basis for an analysis of risks associated with hazards by measuring the likelihood and consequence of hazards in the community. How one perceives the adverse impacts of risk, either from an individual, organizational, or societal perspective, certainly influences strategies to address risk of natural hazards. To say the least, how risk may be perceived and the process for analysis will shape individual and institutional approaches to deal with risk. In this book, we acknowledge that individuals, organizations, and public policy positions may be viewed differently. We stress that an open analysis of hazards is constructive in preparing sound hazard risk management and community hazard mitigation.

Risks may be viewed as voluntary, where we agree to participate in activities that increase our chances of harm or injury, including driving fast or participating in high-risk sporting events. Other risks that we do not choose to participate in are classified as involuntary risks, where we unknowingly or unwillingly are exposed to harm. For involuntary risks, we may be exposed simply because the nature of the risk has changed, as in a potential for wildland fire or a hazardous material spill. Unfortunately, we may not appreciate the actual risk from hazards simply because we have adjusted to the threat they present and not examined alternatives that would reduce our vulnerability.

The Royal Society Study Group acknowledges that risk management as a concept involves making decisions concerning risks and that this concerns both hazard identification and risk analysis. Our use of the term risk analysis fits within this context and reflects our determination to understand the likelihood of a hazard event and the consequences of the disaster on a community, region, or for an organization (1992). This definition of risk analysis comprises the identification of the outcomes and estimations of the magnitude of the consequences and the probability of those outcomes. Finally, organizations use the outcomes of risk analysis to determine what is an acceptable level of risk and if anything can be done to reduce the adverse effects of the risk of a specific hazard. The determination of risk reduction measures on an organizational level is viewed as hazards risk management, while this contrasts with broader community approaches to hazard mitigation and building community resiliency.

Quantitative Analysis of Risk

Our view of risk analysis includes both quantitative and qualitative analysis tools. There are predominantly two categories of analysis: quantitative analysis and qualitative analysis. Quantitative analysis uses statistical measures to derive numerical references of risk. Qualitative analysis uses less-defined ways of describing and categorizing the likelihood and consequences of risk. Quantitative analysis uses specific measurable indicators (whether dollars, probability, frequency, or number of injuries/fatalities), while qualitative analysis uses qualifiers to represent a range of possibilities.

Quantitative representation of likelihood may be provided in the form of a frequency measure such as the number of times of occurrence over a chosen timeframe. For example: 3/year, 1/decade, 10/week. An alternative technique of a quantitative measure would be probability, which reflects the same data as frequency, but expresses the outcome as percentage between 0% and 100% as representing the probability of occurrence. For example, a 100-year flood has a 1/100 chance of occurring in any given year, or expressed as a probability of 1% or .01. Qualitative representation of likelihood uses words to describe the chance of occurrence. Each word, or phrase, will have a designated range of possibilities attached to it as illustrated in the categories in Figure 5.1.

	Chance of occurring in a given year
Certain	>99%
Likely	50–99%
Possible	5–49%
Unlikely	2–5%
Rare	1–2%
Extremely rare	<1%

Figure 5.1 Qualitative representation of likelihood.

Individuals determine the risk of a specific hazard by making a judgment among these alternatives. We base these judgments on many factors which could include our recent experience, how hazards have affected others, information provided by the media, and or community meetings that may have addressed potential hazards.

Quantitative analysis of the likelihood component of risk seeks to find the numerical statistical probability of the occurrence of a hazard causing a disaster. These analyses tend to be based upon historical data. A standard numerical measurement for all analyzed hazards must be established. One of the most commonly used quantitative measures of likelihood, and the measure that will be used in this and subsequent sessions, is the number of times a particular hazard causes a disaster per year.

As was true with the likelihood component of risk, the consequences of risk can also be described according to quantitative or qualitative reporting methods. Quantitative representation of consequence can be represented by the number of deaths or injuries or by estimating actual damages from various events.

Qualitative Representation of Consequence

As was true with the qualitative representation of likelihood, words or phrases that have associated meanings are used to describe the effects of a past disaster or the anticipated effects of a future one. These measurements can be assigned to deaths, injuries, or costs (oftentimes, the qualitative measurement of fatalities and injuries are combined). Figure 5.2 provides an illustration of the subjective ranges to help quantify the measurement indicator.

Critical Thinking: We attempt to understand risk using both quantitative and qualitative tools that allow us to examine hazards and their impacts using both the physical and social sciences. We acknowledge that risk is shaped on an individual basis by the individual's familiarity with local hazards (Slovic 1991), but also from elements of local culture, which includes how hazards have been viewed locally over time. What influences your view of risk?

Views of Risk

Rejeski (1993) notes that discussions of risk have included three primary groups including scientists who form their opinions through a rational process, policy

	Injuries	Fatalities
Insignificant	None	None
Minor	Small number First aid treatment required	None
Moderate	Medical treatment required Some hospitalization required	None
Major	Extensive injuries	Some
Catastrophic	Large number of severe injuries*	Significant number of fatalities Some

*EMA, 2000

Figure 5.2 Qualitative consequence indicators.

makers who establish their perspectives based on multiple sources of information, including quantitative and qualitative data, and finally, the public whose perceptions and judgments of risk are formed from their own perspectives in some circumstances despite data provided by the other two major groups. He observes that environmental risks are full of ambiguities that may not be resolved, especially when interested groups have such different perspectives on the issues. He believes that common view of risk can only be obtained when groups agree to share their perceptions and basis for their positions. He stresses that there is a great difference between uncertainty and ambiguity. For ambiguity, there are intrinsic elements of public policy that separate risk management strategies from the risk analysis process. One of the key elements of debates concerning risk and uncertainty is the relative level of trust that is established between the three groups, scientists, policy makers, and the public. The only option that may be possible to obtain any consensus is to encourage a more participatory process and open dialogue. He concludes that the conflict between these groups is characterized by ambiguity and uncertainty and is a lack of consensus over the proper courses of action to understand the risks associated with hazards. GIS provides a tool for examining both hazards and risk. It provides a tool for visualizing the nature and extent of a risk zone. Unfortunately, this tool cannot solve the problem of disagreement, but it may provide those interested in the risks of hazards with the means of building a consensus.

Hazard models and spatial analysis tools are full of spatial uncertainty that may not be resolved. The goal then is to fully clarify the limitations of our hazard models and the data that is used in the spatial analysis of social, economic, and environmental vulnerability. Figure 5.3 provides an estimate of an accidental release of ammonia on a cool cloudy February day at 10 am, wind from the east at 10 mph.

The release occurred near a hospital when a 600-lb tank was dropped from a truck unloading a shipment of various cylinders. The model output provides three estimates of risk using alternative exposure limits of 25 ppm, 150 ppm, and 750 ppm.

Critical Thinking: The question that the scenario in Figure 5.3 presents centers on our risk of harm for a specific exposure limit in a chemical release. The question of risk in this case is not simple and depends on many factors such as where we are in the risk zone (are we close or further away from the actual release), if we are inside a building or are exposed in the outside environment, our individual health and if we suffer from asthma or other breathing handicap, our age and physical size.

The three exposure limits for the scenario in Figure 5.4 were drawn from Emergency Response Planning Guidelines (ERPGs) which are used in the Areal Locations of Hazardous Atmospheres (ALOHA) chemical dispersion model to predict the area where a toxic gas concentration might be high enough to harm people. Three sets of exposure limits were developed by a committee of the American Industrial Hygiene Association for use as planning guidelines, to anticipate human adverse health effects caused by exposure to toxic chemicals.

The three-tiered guidelines have one common denominator, which is a one-hour direct exposure duration. Each guideline identifies the substance, its chemical and structural properties, animal toxicology data, human experience, existing exposure guidelines, the rationale behind the selected value, and a list of references.

The categories as noted in the ALOHA "Help" tool, do not protect everyone, for very sensitive individuals including young children or older adults might suffer adverse

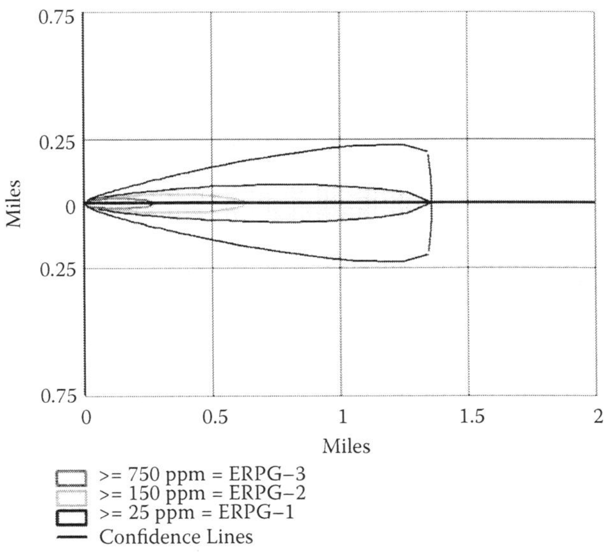

Figure 5.3 (See color insert following page 142.) Hazard risk zones representing alternative exposure limits. Output from EPA ALOHA modeling program.

ERPG 1

The maximum airborne concentration below which it is believed that nearly all individuals could be exposed for up to 1 hour without experiencing anything other than mild transient adverse health effects or perceiving a clearly defined, objectionable odor.

ERPG 2

The maximum airborne concentration below which it is believed that nearly all individuals could be exposed for up to 1 hour without experiencing or developing irreversible or other serious health effects or symptoms which could impair an individual's ability to take protective action.

ERPG 3

The maximum airborne concentration below which it is believed that nearly all individuals could be exposed for up to 1 hour without experiencing or developing life-threatening health effects.

Figure 5.4 Emergency Planning Guide exposure guidelines.

reactions to concentrations far below those suggested in the guidelines. Further, these exposure limits are primarily based on animal studies and not on humans. In addition, the exposure limits are based on a one-hour time period and do not account for any personal safety measures that might be taken to reduce our exposure. The fact is that we might be exposed for a longer period, but seek shelter at the initial signal of the release.

Frosdick (1997) also agrees that our understanding of risk is full of uncertainty and notes that terms such as risk assessment, risk evaluation, and risk analysis are used interchangeably to describe techniques and processes in the management of risk. He notes that, for such a little word, risk is complex and has been the subject of disagreement for some time, especially between natural and social scientists.

Using Historical Data in Determining Risk

One of the largest data sets relating to disasters is maintained by the Centre for Research on the Epidemiology of Disasters (CRED) at the University of Louvain, Belgium, where the Emergency Events Database (EM-DAT) covers both natural and human-caused disasters since 1900. Even with the possibility of a very accurate data set reflecting disasters, how we measure them in terms of losses is complex.

Comparisons between countries can be problematic, for US$1 in one country may have a different value in another. Our ability to measure disasters over time has changed in the methods that we use to collect the frequency of disasters worldwide. Our capacity to detect and accurately classify disasters since satellites have been in use means that our data since the 1960s may be far more accurate than frequency data sets of the early twentieth century. As an example, our ability to accurately detect and classify earthquakes or tropical cyclones is far greater today than ever before. We thus see in many of the data sets a dramatic increase in disasters in the last twenty years.

Numerous data sets reflecting the frequency of disasters and their consequences worldwide are available from the United Nations, Munich Reinsurance Company. These data sets may be of value in establishing a benchmark for a specific type of hazard, which may be adjusted for a specific part of the world. Data obtained from more domestic sources such as the National Weather Service (NWS) or National Oceanographic and Atmospheric Administration (NOAA) may provide a more accurate determination of specific risks of hazards in a part of the country.

For the United States, the National Climatic Data Center (NCDC) serves as a national resource for climate information. NCDC can provide historical data to help document historical climate information. As a climate resource, the NCDC works with scientists and researchers worldwide. They provide both national and global data sets for weather and climate information.

In addition to the NDCD, the USGS Center for the Integration of Natural Disaster Information is a clearinghouse for disaster information and provides links to disaster data distributed by other agencies (Thomas 2001). The U.S. Environmental Protection Agency and the U.S. Department of Transportation (DOT) provide information on accidental releases of hazardous chemicals. DOT focuses its data set on transportation accidental releases, while EPA does fixed site releases.

Thomas notes that, although there has been some integration of hazard event data within a single agency such as the NWS, "a true systematic integration of multiple types of hazard data currently does not exist" (2001: 64).

The Need for Complete Accurate Data for Decision Making

In order to reduce the adverse impacts of disasters, those involved in the hazards analysis process must have accurate and timely information to support effective decision making. Information that results from our hazard modeling exists to support decision making. The sources of this information must be known and shared with those who use our recommendations and hazard analysis outputs. Rosenthal and Kouzmin stress that organizations fully understand the threat imposed by hazards and can establish a framework from which to deal with disaster outcomes (1997).

Quality information is essential in any hazards analysis effort, and in many cases this information is in the form of a technical report and utilizes complex scientific hazard modeling. It is critical that this complex information be used

Hazard	Agency	Time Covered
Tornadoes	Storm Prediction Center Norman, IL	1959-present
Thunderstorm, wind	Storm Prediction Center Norman, IL	1959-present
Hail	Storm Prediction Center Norman, IL	1959-present
Lightning	National Climatic Data Center Asheville, NC	1959-present
Storm data*	National Climatic Data Center Asheville, NC	1959-present
Hurricanes	National Hurricane Center Colorado State University	1886–1996
Floods	National Weather Service	1903-present
Earthquakes	Council of National Seismic Systems	1970-present
	National Geophysical Data Center	2150 B.C.-1994
	Earthquake Research Institute University of Tokyo, Japan	3000 B.C.-1994
Volcanoes	Global Volcanism Program Smithsonian Institution	8000 B.C.-present

*Meteorological events including wind, hail, lightning, water hazards, tornadoes, flooding, drought, landslides, hurricanes, wildfires, and thunderstorms

Figure 5.5 Natural hazard data sources with time covered.

to protect public property and lives of citizens. Data used in a hazards analysis have time and spatial characteristics (Figure 5.5). The data requirements for supporting the emergency management process will vary both for the type of hazard as well as how the outputs will be utilized in supporting decision making (Cutter 2001).

Using Technical Data in Decision Making

The description and categorization of hazard areas, critical infrastructure, and disaster zones is greatly facilitated by the use of geospatial technologies and hazard models. The use of scientific data from hazard models and risk analysis requires that decision makers fully understand the limitations of these tools and how to communicate information. An informed user of complex data is critical to minimizing legal challenges and suits. Hazard models can provide different results with just minimal changes in data inputs. Clarifying the sensitivity to the models and the limitations of data inputs will help to avoid challenges to the use of these models in emergency management.

There may be a discrepancy between an objective assessment of risk by the hazards analysis team and the public (Kirkwood 1994). Clearly an objective view of risk by a knowledgeable professional, who understands the nature and limitations of hazard modeling and how it is described, may not be shared by the public. An objective evaluation of risk must be nonjudgmental and explained in a way that the public or other stakeholders can understand. Kirkwood stresses that, regardless of our scientific modeling, there will be in the eyes of the public some level of risk, however small.

Wallace and Balogh (1985) stress the need for decision support systems (DSS) for using technical data. They stress that a DSS must provide support to decision makers and their stakeholders, evolve as the users become more familiar with the technology, be interactive and controllable, recognize their nonroutine but consequential use, and adapt to the idiosyncrasies that are inherent in human decision making.

Indicators of Direct and Indirect Losses

We measure the consequences of disasters using indicators of disaster impacts. They could include social disruption, economic disruption, or environmental impacts. Social disruption could include the number of people displaced or made homeless, or incident rates of crime (murders, arrests for civil disorder, or fighting); economic disruption could be based on unemployment rates, days of work lost, production volume lost, or decrease in sales or tax income. Environmental impacts could be valued at total cleanup costs, cost of repair or restoration of water or sewerage systems, number of days of unhealthy air, or the number of warnings involving fish consumption or restricting recreational use of a water feature.

Direct tangible losses such as fatalities, injuries, cost of repair, loss of inventory, response costs by a business or community, or relocation costs are first-order consequences, which occur immediately after an event (Smith 2004). Indirect losses associated with a disaster evolve after the event and include loss of income by displaced employees, sales that did not occur, increased costs for skilled employees, losses in productivity of employees, employee sickness, increases in disruptive behavior (fights) at schools, or crime in a neighborhood.

Critical Thinking: How might you measure the intangible losses related to the impact of a disaster in an education system that has to accommodate an increase of 25% more students, or increases in traffic in a community that absorbed 40% more residents who have been displaced?

The indicators for social, economic, or environmental impacts may be based on historical data and collection of data after a disaster event or modeling techniques. To estimate the number of deaths and injuries using historical data, one would determine for past disasters the number of injuries, fatalities, displaced persons, or number sheltered or left unemployed. The population of the community would need to be determined at the time of the event and then the same data collected after a similar disaster event. A rate comparing the number of injuries for the total population would provide a means of comparing injuries at different disaster events, allowing for population changes over time.

Allowing for population changes over time does not account for other related changes that could impact injuries, fatalities, and indicators for disaster impacts. The impacts from more recent disaster events will not be able to account for legal changes (code requirements or flood plain management programs), changes in development patterns, or cultural and social changes causing movement in populated areas.

The use of measurable indicators to help understand risks could be enhanced if all the indicators used the same reference points. An example would be to quantify deaths, injuries, and damages in a common measure such as U.S. dollars. Unfortunately, it may be impossible to associate a dollar amount to some indicators. The alternative is to use measurable indicators that may be compared over time.

The World Bank classifies each national economy by its gross national income (GNI) per capita, to reveal low-income, middle-income, and high-income countries (ISDR Secretariat 2003). A more complex ranking measure of vulnerability is provided by the United Nations in its Human Development Index (HDI), which uses life expectancy, educational attainment, and income as indicators of sustainability.

Intangible losses are those that cannot be expressed in universally accepted financial terms and are not generally included in damage assessments or predictions. Despite the difficulty in associating some intangible losses to specific indicators, we may want to identify some type of indicator that reflects the losses associated with cultural changes, individual and family stress, mental illness, sentimental value, and environmental losses. We need to identify appropriate measures of both tangible and intangible losses associated with disasters. It is not uncommon for the intangible impacts to exceed the tangible ones in terms of the overall effect they have on a community (UNDP 2006).

As we examine potential losses from disasters, we may find that the community or business organization actually has gains. Though it is extremely rare for gains to be included in the assessment of past disasters or the prediction of future ones, it is undeniable that benefits can exist in the aftermath of disaster events. Gains could be observed in amount of employment, business volume, tax collections, number of residents, traffic counts, or crime. Post-Hurricane Katrina data show that many cities within a 100-mile distance of the City of New Orleans had positive gains from the displacement of the metropolitan New Orleans population. Although the impacts were temporary, some gains remained even years after the disaster.

Issues in Risk Analysis

Changes in Disaster Frequency

Changes in disaster frequency may be the natural result of climate variations which occur over a long period or from changes in variables that impact the frequency or severity of an environmental change such as an increase in human activity where the hazard already existed. The number of hurricanes that enter the Gulf of Mexico states varies over a long period, especially with the rise or fall of sea surface temperatures or wind patterns from an El Niño. Flooding or hurricane storm surge might cause more physical damage because of increases in development or in coastal areas where major changes in development have occurred (Smith 2004). Worldwide trends in population shifts to high hazard coastal zones will likely result in higher losses from tropical cyclones. Environmental changes resulting in natural system degradation may also increase the severity of hazards. As more buildings, technology, infrastructure, and other structures are constructed, the potential for hazard impacts increases.

With changes in technology, people expect to have a certain level of services, including availability of water, electricity, and easy long-distance transportation. As these systems expand and develop, they become vulnerable to hazards. Major blackouts, the spread of computer viruses, or communication of terrorist threats have occurred worldwide in the past and will likely occur in the future. The interdependence of our societies globally makes us increasingly vulnerable to epidemics or disease or massive economic changes in stock markets, oil, water, or other natural resources. Disaster trends may need to be examined over very long periods, and unfortunately, we may not have access to data sets to help us understand the natural variation of a specific hazard.

Availability of Essential Data

The availability of essential data for modeling hazards and determining the frequency of occurrence is critical in a valid risk analysis. The following example concerns the availability of historical hydrological data and illustrates the challenge that we face when attempting to understand the nature of the risk presented by a natural hazard.

Hydrology is the science that deals with the properties, distribution, discharge, and circulation of water on the surface of the land, in the soil, and in underlying rocks. It also refers to the flow and behavior of rivers and streams. All flood modeling programs that are used to create flood maps for local communities need a discharge value for a water feature. The discharge value is determined by measuring flow rates directly by the USGS through a river gauge or indirectly by statistical methods. A USGS River Gauge Station measures a water feature discharge for a particular site on a stream, canal, lake, or reservoir where systematic observations of water flow characteristics are obtained. For these stations, data is collected on a real-time basis, since it is data collected by automated instrumentation analyzed quickly enough to influence a decision that affects the monitored system.

Discharge characterizes the volume of water passing a point of a stream or river (river gauge station) or hydrologic unit per unit of time, commonly expressed in cubic feet per second, million gallons per day, or gallons per minute. A major input to a flood modeling program is the discharge value or volume of water in a river or stream.

The measurement of high water is a critical part of the modeling process for flooding events. A hydrograph is a graph showing the variation of water elevation, velocity, stream flow, or other property of water with respect to time. A hydrograph is provided for each USGS River Gauge to describe the properties of the water at that point in a water feature. It provides real-time and historical values. A hydrograph shows a mean discharge (MEAN) of a stream during a specific period, usually daily, monthly, or annually. Figure 5.6 provides an example of a hydrograph for a water feature that has a USGS river gauge. For examples of real-time hydrographs, see the following USGS Internet site: http://cfpub.epa.gov/surf/locate/index.cfm.

The National Flood Insurance Program (NFIP) was established in 1968 and made affordable flood insurance rates available to individuals through participating local communities such as towns, cities, counties, or parishes. In 1983, a common standard for risk assessment and management of flood hazards was adopted by federal agencies and known as the 100-year or 1% annual chance of flood as the

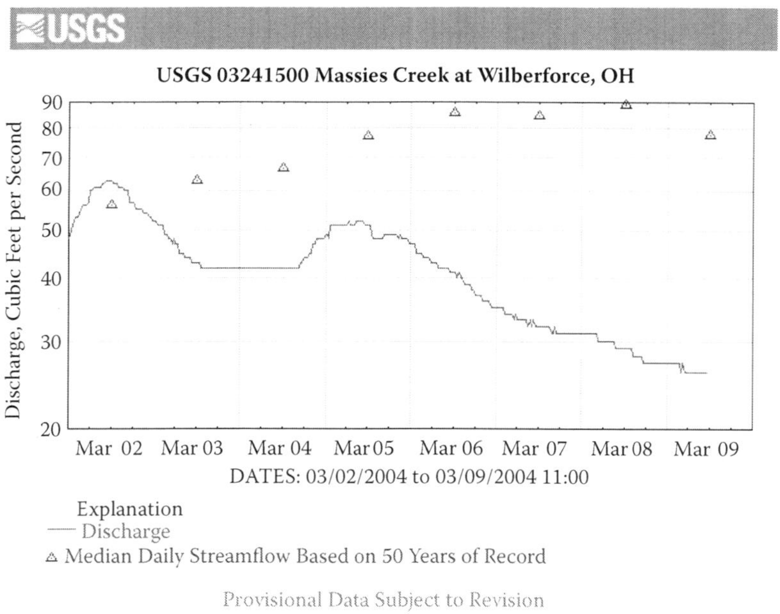

Figure 5.6 (See color insert following page 142.) USGS hydrograph for a water feature.

standard for floodplain management. This standard was considered to represent a degree of risk and damage worth protecting against, but was not considered to impose stringent requirements or burdens of excessive cost on property owners. The 100-year event standard represents a compromise between minor floods and the greatest flood likely to occur in a given area, that the highest recorded flood level reflects what has happened rather than what could happen, and that in many cases the 100-year flood level is less than the flood of record. This 1% annual chance flood (or 100-year flood) and the associate floodplain have been widely adopted as the common design and regulatory standard in the United States. This 1% annual chance flood is used by the NFIP as the standard for floodplain management and to determine the need for flood insurance. The development of flood models and flood maps was considered by the NFIP as a primary means of reducing flood hazards. The flood maps would provide a basis for managing the development and use of flood-prone areas and lead to a better understanding of the magnitude and likelihood of large flows.

Information on water feature flow frequencies would grow in importance as federal agencies enhanced their efforts to assist in flood mapping for communities joining the NFIP. The U.S. Water Resources Council (USWRC) published estimates of flood flow frequencies, (Benson 1967; Ries and Crouse 2002). The techniques were adopted by USWRC for use in all federal planning involving water and related land resources.

Practically all government agencies undertaking floodplain mapping studies use flood flow frequencies as a basis for their efforts (IACWD 1982). Flood flow frequencies are used to determine flood discharges for evaluating flood hazards for the NFIP. Flood discharge values are a critical element in preparing Flood Insurance Rate Maps. Flood flow frequencies may be determined using a minimum of ten years of historical data from USGS stream gauges (Figure 5.7). This data is used in flood models approved for creating flood maps for local communities participating in the NFIP. For water features where USGS stream gauges are not available, flood flow frequencies are based on calculations from the USWRC guidance. The flood flow frequencies using these techniques have been determined for large geographic regions and are statistical representations rather than averages based on data from local water features. Flood modeling based on these statistical representations is considered valid by the NFIP, but the studies should acknowledge the source of data for flood flow frequencies. Although valid, they are not based on local data reflecting water feature flood discharges.

Depth of Analysis

Each hazard that is analyzed must be considered according to the range of possible intensities that could be exhibited by the particular hazard. Depending on hazard, we may need to examine it based on its intensity, since the frequency of occurrence is so rare.

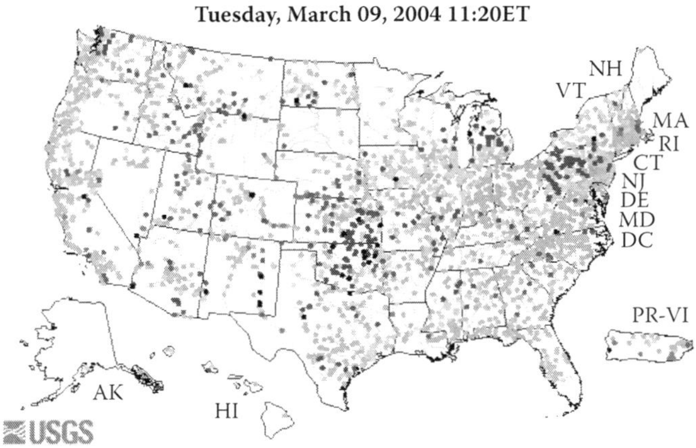

Figure 5.7 (See color insert following page 142.) USGS river gauges in the United States (go to http://water.usgs.gov/waterwatch to review active state stations).

We generally see that lower-intensity hazard events occur more often than more severe ones as in the case of a hurricane or an earthquake. More hazard events, even lower-intensity ones, give us a more comprehensive assessment. We can thus determine a broad-based frequency by calculating the (quantitative) likelihood of each identified hazard (broken down by magnitude or intensity if appropriate); calculating the (quantitative) consequences that are expected to occur for each hazard (broken down by magnitude or intensity if appropriate), in terms of human impacts, environmental impacts, and economic/financial impacts; developing a locally tailored qualitative system for measuring the likelihood and consequence of each hazard identified as threatening the community; and then determining quantitative data into qualitative measures for both the likelihood and consequence of each hazard.

Once a general frequency is determined, we need to estimate damages that might be expected to affect the community. Damages resulting from major disaster events may form the basis for examining the impacts of disasters in future ones. The massive floods of 1993 or hurricanes such as Andrew (1992) or Katrina (2005) could provide a basis for estimating damages in future events. Note that there are inflation converters available to help us determine damages in one time period compared to another.

Losses to structures may be estimated based on total replacement values. The same is also provided for losses to building contents. Losses to structure use and function and cost of displacement may also be estimated for specific critical structures. Although each hazard may affect structures and their contents in different ways, we can estimate the level of potential damage, replacement value, or cost of relocation of operations for that structure from a hazard.

For major weather-related events, the data may be available for large areas. Unfortunately for most hazards, sufficient information does not exist to accurately quantify the likelihood of a future occurrence of a disaster to a high degree of confidence. This is especially true for hazards that occur infrequently and/or occur with no apparent pattern, such as earthquakes, droughts, terrorism, or nuclear accidents.

Dealing with Uncertainty

The very nature of risk analysis is one of uncertainty. Estimates that result from both the quantitative and qualitative analyses are merely estimates, and cannot account for many unknown and confounding variables that affect the way humans and the built environment are affected by disasters of all kinds. Uncertainty may be framed from different perspectives:

- Scientific—precise measurable and available indicators of risk may not be available.
- Legal—potential liability for negligence.
- Moral—individual and societal obligations.
- Societal—personal obligations to others.
- Institutional—system relations between public, private, and nonprofit sectors.
- Proprietary—individual and organizational self-interests (De Marchi and Ravetz 1999).

Because the risk analysis process is so dependent upon timely accurate data, uncertainty is likely to exist in any analyses. Our estimates of risk do involve assumptions and biased judgments that contribute to uncertainty and must be considered when making decisions.

Quantitatively based risk estimates associated with scientific objectivity are unfortunately more subjective than we would want to admit. Our determination of risk is ultimately a very personal, social, and human dynamic rather than one that is objective and neutral. True scientific objectivity is thus a myth and likely unobtainable in our efforts to quantify risk to hazards. Risk thus presents a paradox, for it is just a threat and has not occurred. The risk presented by a hazard is thus an uncertain risk. Since we cannot be certain, our estimates of risk are just a prediction.

Relative Ranking of Risks

Risk evaluation involves the determination of the relative seriousness of the risk of a hazard as it could affect an organization or a local community. Organizations and communities face a range of natural and technological hazards, each of which requires a different strategy to reduce the risk factors of likelihood or consequence.

To facilitate the relative ranking of risks, organizations should determine if a risk may be addressed by another agency, identify which risks require immediate attention, and clarify if the risk associated with a hazard requires further evaluation (Cameron 2002).

We can determine the relative ranking of risks associated with hazards facing our organization or community by considering the following factors:

The likelihood and consequences of the hazard
The voluntary or involuntary nature of the risk (Smith 2004)
Whether there are benefit to cost ratios of mitigating different risks
Political and social ramifications of certain mitigation decisions

The final output of risk evaluation is a prioritized list of risks, which will be used to decide treatment (mitigation) options.

When we examine the likelihood and consequences of a hazard, we use quantitative values. For those hazards that are known to exhibit a range of magnitudes or intensities, these likelihood and consequence values were determined for several of the magnitudes or intensities across the range of possibilities. Assigning these qualitative values was the first step in a process that allows for a direct comparison of the risks faced by a community.

To compare hazards according to their likelihood and consequences, a risk matrix may be constructed. In short, a risk classification categorizes the likelihood and consequence of hazards together in various combinations (Figure 5.8).

An initial risk evaluation of hazards using this approach allows us to classify the risks presented by various hazards. We can then use the FEMA "Multi-Hazard Identification and Risk Assessment" publication to further classify the risks presented by these hazards using the following risk categories.

Extreme Risk: High-risk condition with highest priority for mitigation and contingency planning (immediate action). **Almost Certain Likelihood and Catastrophic Consequence**

High Risk: The likelihood of a risk is high, but the consequence low. Risk addressed immediately by mitigation and emergency preparedness and contingency planning (prompt action). **Almost Certain Likelihood and Minor Consequence**

Moderate Risk: Risk likelihood low, but consequences high. Consideration for mitigation and preparedness critical. **Low Likelihood and Extreme Consequence**

Low Risk: Low-risk condition with additional mitigation contingency planning (advisory in nature). **Low Likelihood and Low Consequence**

Emergency Management Australia (EMA) classifies risks in a similar capacity from Extreme Risk, High Risk, Moderate Risk, and Low Risk (2001). Because a

Hazard	Likelihood	Consequence	Risk
Flood	Possible	major	Extreme
Drought	Likely	Major	Low
Extreme heat	Possible	Moderate	High
Extreme cold	Possible	Moderate	High
Thunderstorm/lightning	Almost certain	Minor	High
Tornadoes	Likely	Major	Extreme
Severe snowstorms	Likely	Moderate	High
Ice storms	Unlikely	Major	High
Land subsidence	Rare	Minor	Low
Earthquake	Rare	Major	High
Transportation accidents	Possible	Catastrophic	Extreme
Hazmat transportation accidents	Unlikely	Moderate	Moderate
Closure of critical transportation routes	Unlikely	Moderate	Moderate
Power failures	Possible	Moderate	High
Water/sewer line failure	Unlikely	Moderate	Moderate
Telecommunications failure	Unlikely	Minor	Low
Computer systems failure	Possible	Minor	Moderate
Gas line break	Unlikely	Minor	Low
Stored chemical leak/accident	Unlikely	Moderate	Moderate
Sabotage/intentional destruction	Possible	Major	Extreme
Biological communicable disease (plague)	Possible	Major	High
Laboratory accidents	Possible	Major	Extreme
Building collapse	Rare	Catastrophic	High
Building fire	Unlikely	Catastrophic	Extreme
Epidemic	Unlikely	Major	High
Widespread poisoning	Unlikely	Major	High

Figure 5.8 Likelihood consequence matrix.

"risk level" may be assigned to more than one matrix box, we will not create an ordered list of risk priorities, but rather several categories of risks with several hazards falling within each category group.

The evaluation can then combine these categories into a spreadsheet to reflect (Lunn 2003):

- Likelihood: Is the hazard likely to occur?
- Consequences: What is the seriousness of the impacts of the hazard?
- Level of risk as determined by evaluation on the risk matrix (Extreme Risk, High Risk, Moderate Risk, and Low Risk).
- Additional considerations including:
  - Other organizations or entities that are impacted by the hazard (potential partnerships, resources, or interdependence of risk management or hazard mitigation strategies).
  - Manageability: Adequacy of existing or potential risk management or hazard mitigation measures or controls.
  - Acceptability: Is the risk acceptable from social, political, economic, or environmental impacts?
  - Change in the risk from the hazard: Will the risk remain the same?

Priority Rating

Once the risk levels of each hazard have been compared to the previously established risk evaluation criteria, the risks must be prioritized or ranked. In risk prioritization, we must consider the degree of control over each risk, the cost, benefits, and opportunities presented by each risk, and decide which risks are unacceptable at any cost.

It is important to note that this summary matrix of risks may change as risk management and hazard mitigation strategies are considered. Thus our risk evaluation provides a better idea of those risks for which mitigation must be conducted at all costs, due to the absolute unacceptability of the risk. However, for risks whose mitigation priority ranks are similar, the factors of cost effectiveness of mitigation, technological availability of mitigation options, and other risk treatment factors will necessitate revisiting this priority list and reranking risks using new information.

Figure 5.9 provides a risk matrix for assessing the likelihood and consequences of risks presented by natural hazards (FEMA 2001).

The following definitions were derived for the risk description categories:

Extreme—High-risk condition with highest priority for mitigation and contingency planning (immediate action)

High—Moderate-to-high-risk condition with risk addressed by mitigation and contingency planning (prompt action)

Moderate—Risk condition sufficiently high to give consideration for further mitigation and planning (planned action)

Low—Low-risk condition with additional mitigation contingency planning (advisory in nature)

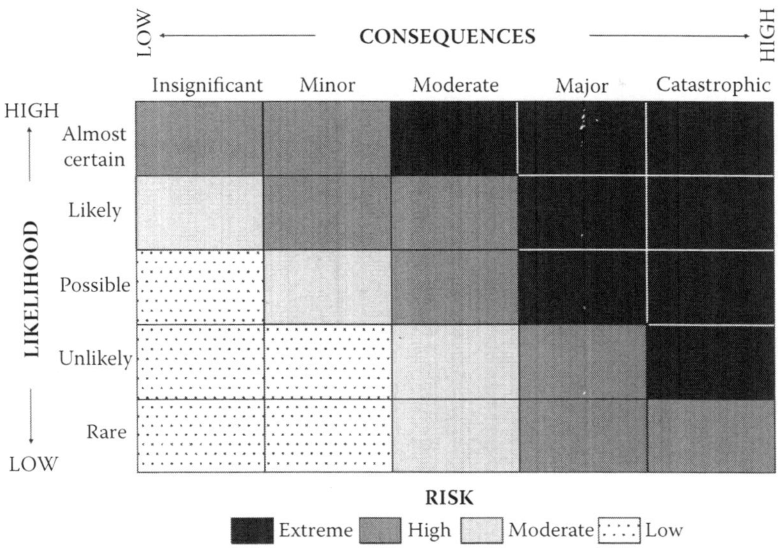

Figure 5.9 Risk matrix. Graphic design by Mary Lee Eggart.

Acceptable Risk

Decisions about what risks to treat, what risks to prevent at all costs, and what risks can be disregarded because of either low consequence, low frequency, or both, must be considered based on our acceptability of risk. The risk analysis process is not working in a vacuum, for there are many factors such as political, social, or economic systems that could affect the determination of what risks are acceptable and what risks are not.

Once hazards have been identified, analyzed, and evaluated, you have a priority list of risks that they must consider for treatment. Ideally, they would treat all risks such that nobody in the community would have to worry about those risks ever again, but that risk-free-world scenario is inconceivable despite modern technology and engineering. While most risks can be reduced by some amount, few can be completely eliminated, and rarely do the funds exist to reduce all risks by an amount that is acceptable to all people in the community.

Risk Benefits

Another factor in the problem of risk acceptability relates to the benefits associated with almost all risks. It is almost universally true that a benefit enjoyed requires some acceptance or tolerance of an associated risk. Living near a water feature may expose you to flooding, and a home in a forest could expose you to a wildfire. To completely eliminate the risk will almost always eliminate associated benefits as well.

When determining the acceptability of risk, we may eliminate certain risks that directly or indirectly establish new ones. For instance, to completely eliminate the risk from nuclear power generation plants, those plants would need to be dismantled and taken out of service. The resulting shortage of power would require that fossil fuel–burning plants increase their production, which in turn would create increased carbon-based pollution, which would likewise create increased health and environmental risks.

Determining Risk Acceptability

There are several factors that together influence the determination of risk acceptability. They include personal, political/social, and economic factors. While the three are very interrelated, they each are driven by different processes. These processes include individual differences, social and political characteristics of a community, and economic considerations.

Personal

Differences in individual acceptance between risks that are voluntary and involuntary in nature are greatly determined by what we see as the benefit from the risk. We are more likely to accept a voluntary risk if we perceive that the benefit is great. Many recreational activities and sports involve considerable levels of personal risk entered into voluntarily. Indeed, the thrill of the risk may be part of the enjoyment of the recreation. When the benefits of a risk outweigh the costs, then the perception of the risk is reduced. In this case the threat level may be considered acceptable. This is especially true when even a high risk is voluntary, as compared to one that is imposed from outside or involuntary. We may tolerate voluntary risks simply because we have learned to adjust to them.

Our personal views of risk and hazards may be what we determine as acceptable risk and associate with our experience with crises and disasters. We may have seen firsthand the potential outcomes from accepting a voluntary risk and believe that the likelihood of harm is so great that we avoid the risk altogether.

In addition to our willingness to accept risk voluntarily, risk may be associated with our individual values, educational experiences, exposure to media coverage of risk, and our individual tolerance of risk.

Political/Social

Because of the differences in the makeup of different communities and populations, risks acceptance will not be universal in all communities and cultures. Risk acceptance is likely to change from place to place, from time to time, and from hazard to hazard (Alesch et al. 2001). Acceptability is likely to change even within individual communities over time as the makeup of that community changes. It is these differences that make the wide public participation in the hazards analysis process so important. Communities that have recently experienced directly the impacts from a disaster will likely be more willing to learn more about a hazard impact and take some type of action.

Economic

Economic considerations of risk are viewed by federal agencies such as the U. S. Army Corps of Engineers from a cost–benefit perspective. The costs of reducing a risk will need to be compared to the benefits (actual risk reduction) that would result. Regulatory agencies such as the U.S. Departments of Energy, Transportation, or Environmental Protection assess risk indirectly, for in most cases private enterprises directly deal with the hazardous substance. Their consideration includes cost factors, but the overall public health and environmental sustainability are higher priorities. Local government agencies have building departments that permit and inspect building codes and safety. Cost benefit is a consideration in the initial adoption of regulations, but extreme events such as hurricanes can motivate public officials to strengthen the codes to protect property from the hazard. Cost considerations may not be the primary driver of the new standards.

Risk Strategies

Accept the Risk

One option is to simply accept the risk, given the present situation and resources of the community. A specific hazard may be considered a low probability of occurrence, and as a result, spending any amount of money would be counterproductive, considering some greater risk reduction that could be gained by using the money to treat another, more severe hazard. The risks that fall within the lowest category of both consequence and likelihood are generally the risks that are considered as acceptable. Members of a community may also believe that the level of risk can be mitigated so as to reduce most adverse consequences. Homeowners who have invested in reducing the vulnerability of their homes and business property may believe that they can withstand a disaster event with limited property damage. They therefore accept the level of risk as they experience it.

Extremely Low Likelihood of Risk

A "de minimis" risk level suggests that the statistical probability of a specific risk from a hazard is so low that concern is not merited. This level is often set at either 1 in 100,000 or 1 in 1,000,000 and is set for a one-year period, or a lifetime (70 years). The term de minimis is a shortened version of the Latin phrase "*de minimis non curat lex*" which means "the law does not care about very small matters." This concept is widely used to set guidelines for levels of risk exposure to the general population such as the chance of personal injury in an airline crash or train derailment or the risk of a reaction from an over-the-counter medicine. For instance, the Environmental Protection Agency (EPA) has set de minimis risk levels for human lifetime risk from pesticides at 1 in 1,000,000 over a 70-year lifetime. The FDA (Food and Drug Administration) and the USDA (U.S. Department of Agriculture) are working on similar regulations of risk for food safety.

Mandatory Risk Level

This type of risk is considered one that is so great that it is mandatory that action be taken to deal with it. It is viewed as an "obnoxious risk" that cannot be ignored, and strategies to reduce vulnerability are mandatory. In practice, this level is generally set at a 1 in 10,000. This practice is often cited in regards to second-hand smoke exposure or accidents in the workplace where safety measures or procedures are required.

Critical Thinking: A tool that can be helpful in understanding the implications of risk is a cost–benefit analysis where we examine alternatives to address a risk including their costs. We then examine the benefits gained by the funds expended. Associated to a cost–benefit analysis is the cost–effectiveness assessment, which examines the minimum unit cost to reduce a maximum level of risk. Consider the risks associated with flooding in a flood zone for an existing older home which is constructed on piers rather than one with a slab on grade foundation. If the unit cost of raising the house is reduced when at least five houses are elevated, then the benefits might outweigh the cost.

Discussion Questions

Would you classify a risk as voluntary or involuntary where factors influencing the frequency and severity of local flooding have changed without the knowledge of the community?

What risks would be included as obnoxious where mandatory action is required?

What risks do you consider to be unacceptable where others have determined that they can live with the hazard? What do personal views, values, beliefs, or your personality contribute to this conclusion associated with a risk?

Applications

Using the risk matrix categories included in Figure 5.8 and Figure 5.9, examine the hazards in your community and categorize their likelihood and damage consequences from low to extreme.

Web Sites

AIHA. Emergency Response Information. The ERPG guidelines are clearly defined and are based on extensive, current data. The rationale for selecting each value is explained, and other pertinent information is also provided. http://www.aiha.org

National Climatic Data Center (NCDC). http://www.ncdc.noaa.gov/oa/about/ncdcwelcome.html

Storm Prediction Center. Storm Prediction Center product and report archive page. NOAA/National Weather Service, Storm Prediction Center. http://www.spc.noaa.gov/archive/index.html

National Hurricane Center, See Hurricane History at http://www.nhc.noaa.gov/pastall.html.

Advanced National Seismic System. ANSS Composite Earthquake Catalog. Advanced National Seismic System, Northern California Earthquake Data Center. http://quake.geo.berkeley.edu/cnss/

Catalog of Significant Earthquakes. National Geophysical Data Center: http://wwwngdc.noaa.gov/seg/hazard/eqint.html

Earthquakes. Earthquake Research Institute, the University of Tokyo, Japan. http://www.eri.u-tokyo.ac.jp/index.html

U.S. EPA. Surf your watershed. Watershed flood monitoring program. http://cfpub.epa.gov/surf/locate/index.cfm

USGS. WaterWatch—Maps and graphs of current water resources conditions. River gauges in the United States for active stations. http://water.usgs.gov/waterwatch

References

Alesch, D. J., J. Holly, E. Mittler, and R. Nagy (2001). *Organizations at Risk: What Happens When Small Businesses and Not-for-Profits Encounter Natural Disasters?* Public Entity Risk Institute, Fairfax, VA.

Benson, M. A. (1967). "Uniform Flood-Frequency Estimating Methods for Federal Agencies." Water Resources Research. U.S. Geological Survey. Vol. 4, #5, pp. 891–908.

Cameron, G. 2002. "Emergency Risk Management: What Does It Mean?" ATEM-AAPPA 2002 Conference, Brisbane, Australia. http://www.aappa.com/infoservices/papers/2002_AAPPA_Brisbane/G_Cameron.pdf.

Cutter, S. L. (2001). "Charting a course for the next two decades," American Hazardscapes: The Regionalization of Hazards and Disasters. Susan L. Cutter Ed., Joseph Henry Press.

De Marchi B. and J. R. Ravetz (1999). Risk management and governance: a post-normal science approach. *Futures* 31(7):743–757.

Emergency Management Australia (2001). Decision Making Under Uncertainty in the Emergency Management Context. Workshop held at the Australian Emergency Management Institute, August 27–28. http://www.ema.gov.au/agd/ema/rwpattach.nsf/viewasattachmentPersonal/4B0DF4E7C3945DB0CA256C85001E2B56/$file/DECISION_MAKING_UNDER%20UNCERTAINTY_IN_THE_EM_CONTEXT.PDF.

FEMA (1997). Multi-Hazard Identification and Risk Assessment. Federal Emergency Management Agency, Washington, DC.

FEMA (2001). Understanding Your Risks: Identifying Hazards and Estimating Losses. Federal Emergency Management Agency, Washington, DC.

FEMA (2002). Mitigation Planning Workshop for Local Communities. Instructor Guide. DMA 2000 State and Local Plan Interim Criteria. Federal Emergency Management Agency, Washington, DC. http://www.fema.gov/fima/planning_toc4.shtm.

Frosdick, S. (1997). The techniques of risk analysis are insufficient in themselves. *Disaster Prevention and Management* 6(3):165–177.

IACWD (1982). Guidelines for Determining Flood Flow Frequency. Bulletin 17B of the Hydrology Subcommittee, Department of Interior, U.S. Geological Survey. Interagency Advisory Committee on Water Data, Office of Water Data Coordination. Reston, VA.

ISDR Secretariat (2003). Living with Risk: Turning the tide on disasters towards sustainable development. United Nations, International Strategy for Disaster Reduction, Geneva.

Jardine, C. G., and S. E. Hrudey (1997). Mixed messages in risk communication. *Risk Analysis* 17(4):489–498.

Kirkwood, A. S. (1994). Why do we worry when scientists say there is no risk? *Disaster Prevention and Management, An International Journal* 3(2):15–22.

Lunn, J. (2003). Community consultation: the foundation of effective risk management. *Journal of Emergency Management* 1(1):39–48.

Millennium Development Goals (2007). United Nations Development Programme. United Nations, New York.

Rejeski, D. (1993). GIS and risk: A three-culture problem. In *Environmental Modeling with GIS*. eds. M. F. Goodchild, B. O. Parks, and L. T. Steyaert. Oxford University Press, New York.

Ries, K. G., III and M. Y. Crouse (2002). The National Flood Frequency Program, Version 3: A Computer Program for Estimating Magnitude and Frequency of Floods for Ungauged Sites, 2002: U.S. Geological Survey Water-Resources Investigations Report 02–4168.

Rosenthal, U. and A. Kouzmin (1997). Crises and crisis management: toward comprehensive government decision making. *Journal of Public Administration Research and Theory: J-PART* 7(2): 277–304.

Royal Society (1993). Risk Assessment, Report of a Royal Society Study Group. The Royal Society, London.

Slovic, P. (1991). Beyond numbers: a broader perspective on risk perception and risk communication. In Mayo, D. and Hollander, R. (Eds.) Acceptable Evidence: Science and Values in Risk Management, Oxford University Press: Oxford, 48–65.

Smith, K. (2004). *Environmental Hazards: Assessing Risk and Reducing Vulnerability*. Routledge, New York

Thomas, D. S. K. (2001). Data, data everywhere, but can we really use them? In *American Hazardscapes: The Regionalization of Hazards and Disasters,* ed. Cutter, S.L. Joseph Henry Press, Washington, D.C.

United Nations (2006). The Millennium Development Goals Report. Department of Economic and Social Affairs. New York. http://mdgs.un.org/unsd/mdg/Default.aspx.

Wallace, W. and F. Balogh (1985). Decision Support Systems for Disaster Management. Special Issue — *Public Administration Review*: 134.

Waugh, W. (2000). Living with Hazards, Dealing with Disasters: An Introduction to Emergency Management. M. E. Sharp Publishers: Armonk, NY.

Chapter 6

Social, Economic, and Ecological Vulnerability

Objectives

1. Define the concept of vulnerability and extreme events.
2. Examine the three dimensions of vulnerability.
3. Clarify how we can measure the three dimensions of vulnerability.
4. Identify examples of the three dimensions of vulnerability.

Key Terms

Vulnerability
Resiliency
Geo-hazard/extreme events
Social system
Infrastructural system
Economic system
Ecological system
Smart growth
Coastal sprawl

Critical Thinking: Why do some communities bounce back and even prosper from disasters while others take much longer to recover or experience delays in restoring their ecological, social, or economic systems? How can we measure the potential impacts from natural hazards on social-cultural, economic, or ecological systems? How can we better understand how these interconnected systems within a community might be impacted?

Introduction

Vulnerability refers to the susceptibility or potential for harm to social, infrastructural, economic, and ecological systems. It is the result of a set of conditions and processes that influence the way that these systems are harmed by natural and technological hazards or extreme events. Vulnerability is closely associated with resilience, which involves the capacity of these systems to bounce back from disasters or their capacity to both respond to and cope with extreme hazard events. We earlier noted that risk was the result of hazard potential, time, and vulnerability. Thus vulnerability becomes central in understanding how our communities deal with risks associated with disasters. Expressed in a different way, vulnerability is the result of our exposure to hazards and our capacity to cope and recover in a sustainable manner.

Approaches to Vulnerability

Recent literature suggests that vulnerability takes many forms, and scholars have developed many techniques that analyze this phenomenon. Three popular methods include the utilization of an exposure model that emphasizes the identification of conditions that make people and places vulnerable to disastrous conditions and is related to the relative frequency and intensity of the hazard, risk, or threat. An exposure model would also allow testing of the vulnerability of critical infrastructure and facilities to impacts of hazardous events. Quantitative approaches in the engineering sciences attempt to assess the infrastructure resilience with the goal to reduce losses through research and the application of advanced technologies that improve engineering, pre-event planning and post-event recovery strategies (Bruneau et al. 2003). Vulnerability as a hazard exposure includes the distribution of people, economies, and the environment to hazardous conditions. The emphasis here is on the physical occupation of areas that may be prone to hazardous events. Under this view, vulnerability is a result of a physical condition that is associated with place (Cutter 1996).

A second approach views vulnerability as a social condition that measures societal resistance or resilience to hazards (Blaikie et al. 1994; Hewitt 1997).

Vulnerability is an outcome of the relation between a hazard and a social condition that includes the capacity to respond and cope in a positive manner. This coping capacity is thus inherent within the resilience of families and groups of people in an area of a hazardous event. The final element is the community's capacity to respond and cope within a geographic area. What resources does the community have to deal with disasters?

Vulnerability includes the robustness of social networks in a community, the strength of critical infrastructure to hazards, an area's risk of a hazardous event, and efforts by the community to reduce potential losses or to mitigate exposure. A community's vulnerability is thus filtered through their social fabric, their efforts to strengthen their infrastructure, and business enterprises' initiatives to reduce their exposure and increase capacity to deal with disasters (risk management). Both community mitigation activities as well as organizational risk management initiatives thus impact the social, economic, and ecological exposure to hazards. Vulnerability is thus more complex than just the exposure of people to hazards, for their efforts to prepare and cope, along with community and organization initiatives, contribute to the community's capacity to respond and cope with disaster events. The fact is that the natural, economic, and social systems are deeply integrated and interdependent in many ways that must be considered in understanding that some communities, people and natural environments are better able to cope and recover from disasters than others.

The third is an integrated approach that examines potential exposures and social resilience (Cutter 1996; Cutter et al. 2000; Kasperson et al. 1988). The integrated approach combines vulnerability associated with risk and exposure with vulnerability as a social response, along with vulnerability of place. Cutter (1996) notes problematic issues even in this integrated approach, because of its lack of consideration of the underlying causes of social vulnerability and its failure to consider distinct spatial outcomes that may vary over time. The variability of risk over a geographic area is central to Cutter et al.'s hazards of place model (2003). Social and biophysical conditions thus interact to produce an overall place vulnerability.

Vulnerability, however, is more than just exposure to and the impact of hazards on essential characteristics of a community's social, economic, and ecological systems. It requires coping strategies by individuals or agencies at multiple spatial and temporal scales. Figure 6.1 shows a conceptual view of vulnerability.

Critical Thinking: The scale at which vulnerability of place is examined may vary from large regions such as metropolitan areas to the neighborhood level. The analysis of vulnerability at the neighborhood level is present in isolated disaster case studies and not included in assessments of large-scale disasters in the United States or internationally. Tornadoes impacting a small community or neighborhood would serve as a very different scale to a disaster such as a hurricane or many floods. What other examples of small-scale disasters can you provide?

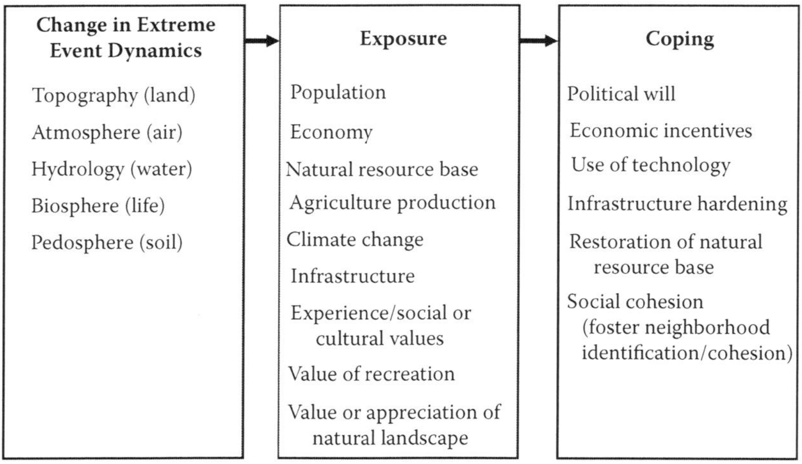

Figure 6.1 Conceptual view of vulnerability. Graphic design by Mary Lee Eggart.

Coping strategies should be included in an integrated approach to vulnerability approach. Vulnerability thus is integrated into the development of action or coping strategies that can be implemented. These strategies reflect choices or public policies that are made by individuals, families, businesses, and public agencies and models that allow testing them. We are interested in who lives in the community and where they reside, but it is the decisions that people make on an individual and collective basis that really drive vulnerability.

For example, some communities may adopt land-use planning and hazard-resistant building codes. For these communities, structures built to the code may be hazard resistant and less likely to be damaged by high wind, floods, storm surge, fire, or other anticipated hazards. Zoning restrictions control building in high-hazard zones either by requiring base elevation for a structure above a specific flood height or by setback requirements from coastal zones. Human choices and our policies are part of our examination of vulnerability.

Vulnerability is influenced by and dependent on coping capacity, so the level of response and recovery can be measured by monetary resources available, deployment of technology by type, resilience of infrastructure, and capacity of the emergency response system. Per capita income may suggest many things about a community and resilience. The United Nations Development Program (UNDP) (2000) sees a relationship between per capita income and fatalities by country. Others see a relationship between per capita income and health attainment (level of life expectancy for a country) as measured by UNDP (2000). Per capita income and (unrestricted) access to medical facilities and health care result in a more resilient population against various disease and therefore longer life expectancy. Economic capacity also provides a base for people in a country to deal with disaster losses—more financial resources mean that countries can cope more effectively. RADIUS (Risk Assessment Tool for Diagnosis of Urban Areas against Seismic Disasters) used hazard exposure, context vulnerability, and emergency response plan (coping mechanism) to shape their indicators of vulnerability (Morrow 1999).

Dimensions of Vulnerability

Vulnerability consists of three dimensions, including social, economic, and ecological elements of our communities. Our goal is to identify sensitive indicators in each area so as to understand how a community might be harmed in a disaster.

Social, economic, and ecologic indicators emerged independently during the 1960s and 1970s specifically designed to provide indices of exposure and environmental health (Cutter et al. 2003). The UNDP has used socioeconomic indicators to examine social and economic implications of regional partnerships (UNDP 2005). The Coastal Risk Atlas is one of the few attempts to link physical hazards and social vulnerabilities (Boyd et al. 2005). Richmond (2001) concluded "there exists no established methodology for determining the hazardous nature of a coastline," and Cutter et al. reconfirmed that metric standards do not exist to assess the vulnerability to environmental hazards (2003). Richmond et al. quantified the effects of only physical hazards to the Hawaiian Islands by historical records and a ranking scheme based on hazard dynamic and frequency to define an overall hazard assessment to be used for coastal land-use planning (Cutter et al. 2003).

Nakagawa and Shaw (2004) note that there are common features that suggest why some communities are more resilient than others. They see that there is a complex mixture of social, economic, religious, and political factors present that influence community resilience to disasters.

Environmental degradation can result in health and economic losses, poverty, loss of intellectual property rights, loss of natural heritage, and conflict exposure to extreme events. It also might be related to the root causes of a hazard outcome such as disease. As an example, water supply, air pollution (indoor), and sanitation are all related to the highest level of risk from disease. This would suggest that indicators are thus related to specific hazards and may be a strong association to some threats while not to others.

Critical Thinking: It is widely agreed upon that social vulnerability is influenced by a lack of information, political representation, richer social networks, culture, infrastructure, age, gender, race, and socioeconomic status, language, and disabilities (Cutter et al. 2003). Hazard potential, geography, and infrastructure conditions interface with the social and economic fabric of a region to influence risk (Cutter et al. 2003).

Social and Human Vulnerability

The social dimension of vulnerability arises from the exposure of people, neighborhoods, cities and rural populations and their capacity to recover from hazard events. The hazards literature has noted that the poor, unemployed, single head of a household, elderly, handicapped, or carless households (Blaikie et al. 1994; Yohe and Tol 2001) are much more likely to suffer the hardest and have more difficulty in restructuring their lives after a disaster than other households that have more resources. The more vulnerable populations take more time than their counterparts to recover following a disaster and as a result suffer to a greater extent. Vulnerability also impacts individual self-protection actions and access to political networks and institutions. Cutter et al. acknowledge these factors, but stresses the geographic dimensions of vulnerability noting that place matters (2003). Too often, the poor and most vulnerable populations reside in the most hazardous zones in a community.

Social vulnerability suggests a differential capacity of groups and individuals in dealing with the adverse effects of hazards based on their positions within the physical and social world (Dow 1992). Historical, cultural, social, and economic processes shape an individual's or social group's coping capacity (Blaikie et al. 1994). Research studies suggest that specific populations are far more vulnerable to the risks from natural and human-caused disasters (Cutter et al. 2003; Peacock et al. 2000). These studies also indicate that there is a strong relationship between socioeconomic vulnerability and disasters and that social and economic costs of disasters fall unevenly on these population groups (e.g., Blaikie et al. 1994; Bolin and Stanford 1991; Cutter et al. 2003; Heinz 2000; Mileti 1999; Morrow 1999).

It is widely agreed upon that social vulnerability is influenced by a lack of information, political representation, richer social networks, culture, infrastructure, age, gender, race, and socioeconomic status, non-English speaking, and disabilities (Cutter et al. 2003). More valuable homes and higher incomes increase resilience to hazards and reduce risks (Cutter et al. 2000). Thus, hazard potential, geography, and infrastructure conditions interface with the social and economic fabric of a region to influence risk (Cutter et al. 2003). The key question raised by these studies centers on the suggestion that some groups are at greater risk than others.

Carter (2006) takes a different perspective on social vulnerability, observing that for some, droughts, hurricanes, and other environmental disasters deal a blow to the poor and vulnerable populations in many parts of the world, so as to trap them in poverty, despair, and dependency. They view patterns around the world to suggest that the poorest households struggle to overcome the desperate situation that disaster or shocks deal them. Their short- and long-term well-being and sustainability make it impossible to ever catch up with wealthier households.

A hurricane hazard vulnerability assessment conducted during 2005 for the Mississippi Gulf Coast combined a GIS-based risk atlas and hurricane simulations (Boyd 2005). Risks were ranked such as flood zones, and vulnerability was examined using income, age, single parents, education, non-English, vehicle ownership, home ownership, and type of home to identify populations at risk and hurricane hazards.

Nakagawa and Shaw (2004) note that there are common features that suggest why some communities are more resilient than others. They see that there is a complex mixture of social, economic, religious, and political factors present that influence community resilience to disasters. They found that the resilience of communities to recover following a disaster is based on both social and economic activities that are heavily influenced by social capital or the level of trust present in the community, social norms, degree of community participation, and finally, the presence of strong community networks.

Critical Thinking: Hoffman (2003) examined who might be hidden victims of disaster and suggests that some very vulnerable people fall through the cracks in disaster recovery, not getting the type of relief needed, and endure ongoing suffering as a result of their situation. She explains that those less able to prepare or cope with disasters are poor or working classes and are some of the most unprotected people in a disaster. As a result of catastrophes, some people slip into a state of perpetual misery. These hidden victims could include undocumented workers, people who lost rental housing (owners or renters) and who did not have insurance, the mentally ill or those with chronic illnesses, people who are severely incapacitated or people who are viewed as social parasites such as beggars, trash scavengers, hustlers, or just the homeless. She raises the question of what happens when those hidden victims who are at the bottom of our society or bottom of the heap are not helped. What happens to the rest of society?

Economic Vulnerability

When we look at economic vulnerability, we examine our risk to changes in the production, distribution, and consumption of goods and services from the private commercial sector but also from the nonprofit and public sectors. The health and vitality of a community's economy is interdependent with the region, nation, and world. The identification of local, regional, national, and international forces that influence local wages, production, export volume, unemployment, and the number and types of jobs may be impacted by many external forces. There are many linkages in our economies that shape the robustness of our local, regional, and state economic base. Suggesting that we can predict accurately how to establish a highly productive economy is very different from the examination of a set of economic indicators that will suggest that a local community could withstand or recover from a natural disaster. Our task then is to identify and examine indicators that will suggest how robust our economy is for a given community and its capacity to contribute in a positive manner to a recovery from a disaster. Economic vulnerability also includes factors that could harm a labor force such as human disease or epidemics.

United Nations World Vulnerability Report (UNDP) documents indicators for indexing and monitoring the potential for disasters.

When we assess the economic vulnerability, we evaluate not only jobs and the nature of the local economy but the capacity of roads, bridges, airports, rail lines, hospitals, prisons, manufacturing plants, shopping areas, utilities, and communication systems to withstand a disaster. It is the potential impact to employee wages, employment, and infrastructure such as electrical, natural gas, and communication sectors that impacts our community's capacity to recover from a disaster. As Comfort et al. (1999) point out, our vast set of services to our rural and urban communities offers a vital backbone to our commerce and standard of living; the scale of these systems also creates dependence and losses that have vast consequences on our economic stability.

The infrastructural and economic vulnerabilities are in fact tightly connected, but can be clearly separated if we consider two aspects: a physical and a nonphysical aspect. While the built environment and its physical resilience against extreme events may be impacted by the physical forces of a hazard, the economic resilience would deal with pressures and impacts of the global economy. In today's global economy, financial, trade, and policy decisions in other parts of the world may have a significant impact on a local economy.

International agencies judge the size and structure of an economy, exposure to international trade shocks, as well as extreme natural events to justify loan or aid programs (USAID 1999). The U.S. AID examines economic vulnerability by determining the frequency and intensity of hazards and conditions such as energy dependency, export characteristics and destinations, and reliance on external financing (Crowards 1999). Munich Re Group (2002) looks at disasters from an economic perspective, including annual per capita income as a reflection of purchasing power. In the agricultural sector of our economy the production of various goods can be measured. But production is highly impacted by external forces such as soil moisture or meteorological forces or geological variables reflecting the hazard itself.

Environmental Vulnerability

Ecological dimensions of vulnerability refers to the capacity of our natural systems to bounce back from disaster. It is the inability of our natural systems to deal with stress that may evolve over time and space (Williams and Kaputska 2000). Saltwater intrusion into freshwater marshes can cause the impairment and even the loss of breeding grounds for fish and other water creatures, birds, and other coastal animals. Long-term intrusion of saltwater into marsh areas can also impact community surface water systems. Hazardous material contamination that results from flooding, wind, or storm surge can cause immediate and long-term decay of delicate coastal environments.

Environmental systems are also significant to the quality of life for a community and its productivity as well as sustainability. Critical views of the rate of deforestation, annual water use as a percentage of total water resources, population density, annual use of water by a household, volume of recycled materials per household, and the relation of coastline to land area. There could be a relationship between number of threatened species in a land area and the ratio of total number of natural disasters to land area (1970–1996) (Parkins 2000). The Yale Center for Environmental Law and Policy (YCELP) identified five components for environmental sustainability including (1) the health of environmental systems, (2) environmental stresses and risks, (3) human vulnerability to environmental impacts, (4) social and institutional capacity, and (5) global stewardship (World Economic Forum 2000).

Lovins et al. (1999) advocates that businesses restore, sustain, and expand our ecosystem so that it can produce vital services and biological resources abundantly. This view suggests that our natural environment as natural capital is to be used, but in a conscious manner so as to reduce waste and expand the productivity of our natural resources. They suggest systems thinking so as to reduce energy costs and waste products. Energy savings can be productivity enhancing. This approach suggests that per capita energy and water consumption is a valid indicator of efficient natural resource allocation and consumption. Further, waste minimization also fits within this model, and thus per capita waste generation and recycling are good indicators of natural systems sustainability. Waste minimization and pollution prevention are also EPA-recommended risk management strategies. Finally, they recommend that we view our natural environment as natural capital and one where we make an investment that will lead to positive return on our investments.

EPA goes further to suggest that a healthy environment provides us with clean air and water, rainfall, productive oceans and water features, fertile soil, and sustainable watersheds. Social economic and environmental sustainability are interdependent, and you cannot have one without the other.

Hossain (2001) noted the efforts of the Commonwealth Secretariat in using indicators to understand environmental sustainability including annual rate of deforestation, population density, and annual water use as a total water resource (Parkins 2000). The World Bank (1999) approach to environmental analysis is based on climate, water, forest, and pollution. Environmental degradation can result in health and economic losses, poverty, loss of intellectual property rights, loss of natural heritage, and conflict exposure to extreme events. It also might be related to the root causes of a hazard outcome such as disease. As an example, water supply, air pollution (indoor), and sanitation are all related to the highest level of risk from disease. This would suggest that indicators are thus related to specific hazards and may be a strong association to some threats, while not to others.

Measuring Vulnerability

An indicator is a quantifiable measurable reflection of a phenomenon. We can use indicators to understand a community's capacity to suffer from, cope with, or recover from a disaster. Indicators are also measurements that help us to understand key assets in our community. By looking at these indicators over time, we can tell if our community is improving, declining, or just remaining the same.

No set of measures tell us everything that we want to know. The Dow Jones Industrial Average does not include all stocks, nor does the Consumer Price Index examine all goods. The key is that we use indicators to give us a barometer of how well something is doing. Quality indicators reflect existing and objective data from well-known sources. The indicators measure something that reflects local conditions or assets that are valued by the community.

Damage measures tangible concrete things that are usually built, such as bridges, homes, commercial or industrial buildings, cars and trucks, or communication towers. Coping indicators demonstrate a community's preparedness for a disaster, such as persons evacuated or sheltered. The number of people evacuated or sheltered could reflect effective warning systems or procedures to help people get out of harm's way. These indicators may reflect a community's preparedness, the effectiveness of hazard mitigation or response strategies, and could explain why some communities recover more quickly than others, suggesting that they are more resilient.

No set of indicators can be all-inclusive. The Dow Jones Industrial Average, a widely respected indicator of stock market performance, does not include every stock traded on the New York Stock Exchange. Nor does the Consumer Price Index measure the prices of all consumer goods. Both indices, like the Sierra Nevada Wealth Index, are based on developing and monitoring a sample of indicators which, viewed together, provide a barometer of overall performance. The 60 indicators serve as an index and were selected because:

- They are measurable and can be updated with existing and objective data sources.
- They measure the condition of assets of material importance to the Sierra Nevada's wealth.
- They measure the condition of assets where active public interest exists.

Indicators of Social Conditions

One of the key issues that we face is to understand if socioeconomic population characteristics indicate higher vulnerability (Cutter et al. 2003). High-risk groups, such as those with lower incomes, the very young and elderly, the disabled, women living alone, female-headed households, families with low ratios of adults to dependents, ethnic minorities, renters, recent residents, tourists, and the homeless, are good social vulnerability indicators of risk and that "social and economic costs of disasters fall unevenly on [these] different classes of victims and stakeholders" (Bolin and Stanford 1991; Cutter et al. 2000; Blaikie et al. 1994; Heinz 2000; Morrow 1999). Economic income limitations impact many people, and thus they are not prepared to deal with or recover from disasters (Mileti 1999). Age (less than 18; over 65), gender (females), race, and income (mean household value) are viewed as primary social vulnerability indicators (Cutter et al. Handbook).

Common indicators that reflect a community's vulnerability include nonwhite population, household incomes less than $25,000, households who rent, number of individuals over 65, disabled individuals (not including employment disabilities), individuals over age 25 without a high school diploma, households without a vehicle, renters, single-parent households with children under the age of 18, and households without a phone. Where the indicator is expressed by household, the value was then divided by the number of total households in the block group.

- Households earning less than $25,000: Percentages of households earning less than $25,000 represents the number of households with earnings less than $25,000 divided by the total number of occupied housing units in the block group. This indicator was chosen to reflect an income threshold instead of households living at or below poverty. This income value reflects the minimum required to qualify for a home mortgage for a home.
- No vehicle: Percentages of households by block group without a vehicle (car, truck, or van) was selected as one of our social vulnerability indicators. The Census Bureau defines housing unit as a house, apartment, mobile home, group of rooms, or a single room that is occupied as separate living quarters.
- Nonwhite race: Much of the hazards vulnerability research has suggested that a nonwhite race is usually located in the highest hazard areas (Bolin 1986; Peacock et al. 2000; Pulido 2000). The percentage of nonwhite population by block group in New Orleans was selected as a social vulnerability indicator and consists of African-Americans as well as Asian and Latino neighborhoods. Howell (2005) noted that African-Americans were less likely to have evacuated or retreated to a safer place for Hurricane Georges and thus could be at a greater risk for Hurricane Katrina. FEMA prioritizes vulnerability in the order of (1) income distribution, (2) elderly populations, (3) disabled populations, (4) children, (5) minority neighborhoods, and (6) language and cultural barriers. FEMA suggests what populations are vulnerable to specific hazards; it is likely that vulnerable populations can be assumed as having similar characteristics.
- The elderly: The percentage of persons within a block group over the age of 65 was selected as a social vulnerability indicator. The hazards literature sites numerous studies that suggest that the elderly are a particularly vulnerable population to hazards. These studies note that physical, mental, and sensory skills become weaker with age. Age is recognized as an indicator of social

vulnerability due to mobility limitations, major dependence on relatives, frequency of respiratory distress, and a lower resilience after the disaster (Cutter et al. 2000; Hewitt 1997; Mileti 1999; O'Brien et al. 1992). Howell found that individuals over the age of 65 were less likely to evacuate or have a plan for evacuation for Hurricane Georges (2005), which threatened New Orleans in the summer of 1998.

- Disability: The percentage of noninstitutionalized individuals with a disability includes persons who have sensory, physical, mental, or other self-care limitations that limit their activities outside the home. This category does not include those with employment disabilities. The percentage is based on the number of disabled individuals in the block group, unlike the other indicators, which use households as the measurement.
- Education: This indicator includes the percentage of individuals over the age of 25 with no high school diploma. Lower education has been suggested as a constraint in understanding hazard warnings (Heinz 2000). The indicator reflects the number of people over the age of 25 in each block group rather than the number of households.
- Use of rental housing: The percentage of rented housing units represents the number of occupied households renting divided by the total number of occupied households in a block group. Several studies suggest that renters are vulnerable because of their lack of finances and/or limitations in transportation (Heinz 2000; Morrow 1999).
- No phone: The percentage of housing units without a phone represents all households without a phone divided by the total occupied housing units. Telephones are an important means of communication to notify people of an immediate evacuation. Calls can be made from emergency managers to inform people of evacuation and locate those who are willing to answer the phone and respond with key punches to indicate if they do not have means to evacuate. Although phone calling as a means of communication has not yet been documented as a warning tool, access to a phone is an important vulnerability criterion. Automated systems for early hazard warning utilize local phone capabilities (Burby 1998).
- Single-parent households: The percentage of single parents who are the head of a household was selected as a social vulnerability indicator. This indicator reflects single parents who are the head of a household with children under the age of 18.
- Number of persons injured or fatalities.
- Number of weeks injured workers out of work.
- Percent of population forced to evacuate.
- Duration of displacement.
- Percent of population below poverty level displaced by the disaster (personal wealth).
- Duration of population below poverty level displaced.

- Crime incidence following a disaster as compared to pre-event levels.
- Percent of population living in a high risk zone (flood zone).
- Percent of high school dropouts (annual basis).
- Percent of students failing high school exit exam.

Indicators of Economic Conditions

Traditional indicators of economic conditions center on employment, housing, business sales, business taxes, and construction. Unemployment rate has been an excellent economic indicator, for it reflects each of the sectors of the economy as well as the public and private employment. Construction has also been used to judge the vitality of a local economy, separating permits for industrial or commercial operations from either rental or home construction. Examining the number of housing starts over a long-term period is a good way to determine if the present condition is performing at a higher rate. Business taxes are also a good indicator of local economic health, with the number of car sales a steady measure of local conditions.

In selecting economic indicators, the measures should be a reflection of the broad basis of the economy and not heavily focused on a single sector. The number of jobs in each of the major economic categories as viewed over a long period provides this type of broad measure. Unemployment rates also provide a broad view of the economy. These indicators include the number of employers and employees (by sector industry), overall unemployment rate (especially small business which has from 1 to 5 employees), the percentage of business failures, household population, and the number of students in schools by type of school or college. Employment estimates provide a good economic indicator for the local economy as reflected in the employment in Figure 6.2 for Orleans Parish following the impacts of Hurricane Katrina and Rita in 2005.

Zandi et al. also provide a breakdown of volume of production as a key indicator of recovery from a disaster, including fishing, chemical production, retail sales, and home and rental prices (average home sale price and the fair market rent for a two-bedroom apartment) (2006). Comparisons of production volume over multiple years is an excellent indicator of the health of the local economy.

Number of residential units destroyed (compare renters and homeowners)
Number of weeks to restore residential units for use
Percent of electrical system shut down
Duration of recovery period to restore utilities
Percent of businesses closed because of the disaster
Percent of businesses open one month or six months after a disaster (recovery period)
Production level of agricultural commodities or units processed by month (comparison of production levels) (single sector or multiple sectors present in a community). To what extent is the community dependent on a single sector?

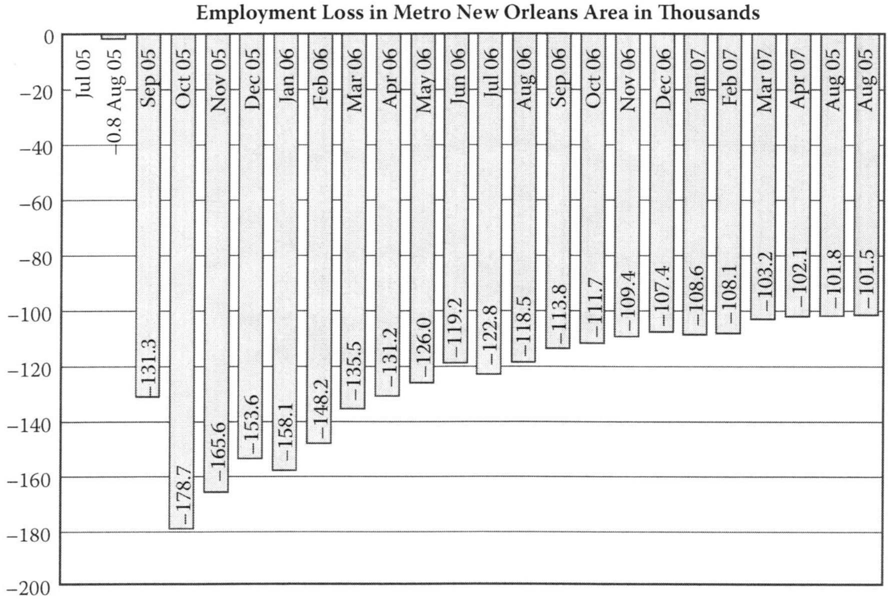

Figure 6.2 Employment loss in Metro New Orleans area in thousands.

Average weekly wage prior to an event and afterwards (by month)
Percent of homes built prior to 1992 (pick a specific data for a community when a building code was adopted)
Percent of residents in a flood zone with NFIP flood insurance
Percent of homes, rental units, or businesses with flood insurance

Indicators of Environmental Conditions

In a study of the effects of urban design on aquatic ecosystems in the United States, Beach (2004) examined the relationship between land use and the effects of sprawl on both air and water quality. He demonstrated that, as the percent of impervious surfaces increased in a coastal community, the nature of the water runoff into water features changed, causing increased levels of nitrogen and phosphorus, organic carbon, trace metals such as copper, zinc, and lead, and pesticides (Schueler and Holland 2000). This dynamic is considered coastal sprawl, which is the expansion of low-density residential and commercial development scattered across large coastal land areas. The result was that changes in urban growth patterns affected habitat quality, water temperature, pollutants, and aquatic life. Further, as coastal communities expanded using traditional development patterns of sprawl, drivers were forced into longer trips for work, recreation, or just normal shopping impacting air quality.

Beach's work centers on a national and even worldwide problem of development patterns in coastal areas. Coastal counties make up 17% of the land area in the United States but just 13% of the nation's acreage. Unfortunately, this coastal zone is home to more than half of the U.S. population. The issue unfortunately is not in the present condition but that this coastal region is where the United States is experiencing population increases. We are continuing to put more people in a small area. Pollution and habitat degradation are the end results of this pattern. The population density of a community would be part of an effort to determine the burden that people have on the environment. This indicator alone does not reflect the magnitude of human impacts on environmental health. We need further indicators.

Beach cited studies demonstrating that when impervious surfaces cover more than 10% of a watershed, water features and estuaries become biologically degraded. A key indicator for a community is the percent of the watershed that is composed of impervious surfaces. When it exceeds 10% there will be problems, according to Beach (2004). The fact is that the ecosystem health, including streams, marshes, and rivers impacted by development, results in less diverse, less stable, and less productive watersheds. Walker (1996) examined two streams and compared the effects of development patterns on the ecosystem health. Increases in impervious surfaces lead to higher levels of sediments containing higher concentrations of nitrogen and phosphorus, organic carbon, or metals such as copper, zinc, or lead, as well as petroleum hydrocarbons and pesticides (Schueler and Holland 2000). The study found differences in channel erosion, but also the health of estuaries. Increased levels of nitrogen lead to algal blooms and fish kills. Increases in fertilizer use in watersheds also reduce water clarity, allowing less light to penetrate below the water's surface, impacting the health of biologic habitats and aquatic habitats.

The issues related to urban development patterns are not limited to the productivity of watersheds but also to the changes in the volume of water in streams or the discharge. The fact is that there is an increased threat of flooding as development increases in a watershed and impervious surfaces increase (Booth 1991; Booth and Reinelt 1993; National Research Council 2000). Changes in water feature discharge reflect development patterns and provide a clear indicator for potential flooding problems in a community.

Changes in impervious surfaces also result in the rise of temperatures in the water. As the percent of impervious surface area increases in a watershed, the water temperature increases (Galli 1991). The result may be decreases in oxygen levels resulting in changes in the marine life and environment.

Hypersprawl is noted as the expansion of residential development with housing densities of one unit on three acres or less. The indicator in this case would be the number of housing units per three acres of land. Measuring housing unit density thus provides us with a means of determining if growth patterns will impact our water features or watershed habitat and should be adjusted to reduce the harmful impacts of urban growth.

Vehicle miles driven by residents also have an impact on environmental health. As the vehicle miles driven per household increase, air quality is impacted. Unfortunately, our urban growth patterns require people to drive further, so that the average commuter trip increases annually. Many communities have adopted more stringent regulations requiring the use of cleaner gasoline or additional measures to prevent pollutants. One can further measure growing emissions by looking at the average interstate highway travel speed. For the United States, the average speed has dropped from 53 to 41 miles per hour, or a 23% decline (Wallis et al. 2001). The result of the decline in the spread is that the average drive takes longer and pollution increases.

EPA has adopted measures to determine when water features might be impacted negatively from agricultural or related practices (use of fertilizers on commercial or residential properties). The measure of Total Maximum Daily Loads (TMDLs) is used by EPA and state regulatory agencies to determine how much pollution a body of water can accept without becoming degraded.

The use of these environmental indicators thus provide a means of assessment and monitoring our environment so as to determine if we are having a positive or negative impact on water or air quality, aquatic habitat, or the risk of potential flooding in the watershed. Our natural environment as reflected in in Figure 6.3 adds beauty to our quality of life but also contributes to a much broader sustainable natural system.

To determine how much pollution a body of water can accept without becoming degraded. Examples of indicators of environmental conditions include:

Population density (high density may reflect vulnerability and exposure to specific risks)
Area contaminated because of hazardous spill and duration of cleanup
Access to transportation infrastructure (access to transportation)
Amount of rainfall per month/annually
Per capita water use
Volume of commercial and or industrial water use
Deforestation

Methodological Issues

When we assess our local economy, as well as our education, public safety, or public health systems, we need to view them from the same scale. That is, we would measure them from a county, city, or neighborhood level. To ensure that any comparisons that are made are valid, we should make sure that our data reflect the same scale. As an example, if we are assessing the capacity of a community to deal with a disaster, we might obtain data on a county-wide basis and look at Census population data, crime rates, educational attainment data, and public health information for the county. We would then compare the result with other surrounding counties,

Figure 6.3 (See color insert following page 142.) Environmental capital: healthy forest, clean water, and soils that support fish and wildlife.

the state and the United States. If we have some data at one scale and then other data at another, it makes our analysis more difficult, including our ability to determine problems that the community might be facing. The key is to ensure that the information that we collect is at the same level (county, census tract, or ZIP code level).

Weighting, Data Availability, and Accuracy

In reviewing the indicators of community vulnerability, a decision would need to be made as to how we view each indicator. Are they of equal weight, or do some indicators have a stronger association between the community's capacity to deal with a disaster than others? Most hazards analysis studies use an equal weighting process where all the indicators are treated the same. Some assign specific weights to various variables (SOPAC 2000).

Throughout this text we have stressed the importance of ensuring that data is current, accurate, and available for use in a hazards analysis. As we examine

Figure 6.4 Healthy landscapes support recreation.

indicators of community vulnerability, the same data quality issues are present and should influence how we judge the capacity of our community to cope with disaster.

Multiyear Analysis

Twigg and Bhatt (1998) contend that vulnerability is complex and too complicated to be represented by models or frameworks. They see that economic, social, demographic, political, and psychological elements influence how communities are impacted by disasters. Vulnerability is too dynamic and complex to account for the many factors that shape our societal adjustments.

The application of quantitative indicators for community resiliency provides a basis for identifying hazards risk management strategies and hazard mitigation alternatives to reduce vulnerability. In addition, qualitative data can be obtained that reflects the views and impressions of community members as to the causes of vulnerability and potential solutions. Surveys of community members can reveal why some citizens refuse to adopt protective actions where notice of a disaster is present. They can clarify what citizens thought would be the result of a specific hazard event and why they refused to evacuate, take shelter, purchase flood insurance, elevate their home, or take other protective measures. Statistical measures of a community can give us clues as to the causes of vulnerability, but only people can explain why they adopted a specific coping strategy and rejected others.

Few studies have assessed the impacts of disasters on communities and examined the relationship between social and economic vulnerability and risk. Hurricane Katrina and the devastating destruction of levees, businesses, and residential property provides an opportunity to examine the relationship between social and economic vulnerability and risk. Did the extent of flooding associated with Hurricane Katrina treat households equally in the City of New Orleans? Did the storm have a far greater physical impact on households that are considered to have a higher social vulnerability?

Interdependence of Social, Economic, and Ecological Capital

Social, economic, and natural systems do not exist independently but in concert with one another. Some of the indicators from one capital might fall into another system; for example, unemployment has both human and economic dimensions; educational attainment impacts both social and economic dimensions. Water quality is an indication of environment capital, but communities where water sports and recreation are a critical economic base certainly fit within the two capitals. Figure 6.4 reflects a healthy natural environment but also contributes to our economic and social capital. The interrelatedness of these systems does not complicate how we define vulnerability indicators, but reflects on the resilience of the community in two ways.

Resilience is the capacity of the community impacted by a disaster to recover quickly. The stronger the social, economic, and environmental capital, the more resources community has from which to draw for recovery. If one conceptualizes the extent of each type of capital matters, for low unemployment the higher the average weekly wage for the community reflects a stronger economy. Further, if there is a low crime rate, high school dropout rate, or lower percentage of households below the poverty rate, one would suggest that the community is in a better position to sustain a disaster. More resources are available to cope with recovery.

If we represent the three types of capital by a circle, then the size of the capital would represent a stronger robust capacity for dealing with disaster. It might also reflect that one type of capital is stronger than another, such as a vital petro chemical industry might have high wages but have low indicators, reflecting the natural environment and higher levels of pollution.

A second view of the relationship between social, economic, and natural environments is their interdependence. A community that relies on the attraction of the natural environment would be drastically impacted by a forest fire, hurricane, flood, earthquake, or other disaster. Since each community is unique in its natural assets, culture, and economic assets, the overlap between them would vary and influence the community's resilience and sustainability. As leaders in the community examine their social, economic, and natural systems, they should not only determine their robustness but also their interrelatedness. Every community is unique, and the interrelatedness of their social, cultural, and natural assets determines the robustness of their assets to deal with disaster.

Discussion Questions

Should the vulnerability indicators be hazard specific or used for any hazard? Should the indicators be selected for a specific geographic area or community?

Are indicators related to specific hazards or do they apply to any hazard?

Applications

Using one of the sources of capital for a community (social-cultural, economic, or environmental), select six indicators that you believe are a broad reflection of the resilience of your community. For the six indicators, determine an appropriate scale as well as a time period for the data. The scale could be by county or city, and the time could be quarter or annual. In addition, determine if your indicator will be expressed in terms of a percentage of households, population, or persons in the workforce. Examine these indicators for your community, and determine if you believe they provide a valid basis for assessing the sustainability of your community.

Web Sites

American Planning Association. APA represents practicing planners, elected and appointed officials, and citizens involved in planning issues. http://www.planning.org

Indicators Sustainable Development. http://www.sdi.gov/

Johns Hopkins Bloomberg School of Public Health. *Chesapeake Bay Indicators Project.* Report Highlights Importance of Chesapeake Bay Health to Human Health. http://www.jhsph.edu/publichealthnews/press_releases/PR_2004/Burke_Bay_indicators.html

National Academy of Sciences (2004). Key National Indicators Initiative. National Academy of Sciences. http://www.keyindicators.org/

NOAA Coastal Services Center. Living on the coast: smart growth tools on the Internet. http://www.csc.noaa.gov/smartgrowth/text/resources.html

Pacific Disaster Center. World natural hazards Website. http://www.pdc.org/iweb/pdchome.html;jsessionid=B8324F79D2423AABC5BF595F8DCDC8CD

Partnership for Sustainable Development. Children's Environmental Health Indicators. http://webapps01.un.org/dsd/partnerships/public/partnerships/206.html

Sierra Business Council. Sierra Nevada Wealth Index 3rd Edition (2006). The Sierra Business Council (SBC) developed the Sierra Nevada Wealth Index to help business leaders and policy makers understand the assets that sustain our region. The Index describes the social, natural and financial capital, which are the foundation of the Sierra Nevada's economy and thereby provides an integrated understanding of our region's wealth. http://www.sbcouncil.org/2006-Wealth-Index

SOPAC (2002). Towards Managing Environmental Vulnerability in Small Island Developing States (SIDS). South Pacific Applied Geoscience Commission. http://www.un.org/special-rep/ohrlls/sid/sid2004/SOPAC-concept%20paper.pdf

Sustainable Communities Network. Smart Growth Online The Smart Growth Network is a coalition of organizations working together to promote smart growth. The Web site features a calendar of events, legislative news, reports, links to partner organizations, and more. NOAA is a partner of the Smart Growth Network. http://www.smartgrowth.org/Default.asp?res=1024

United Nations Department of Economic and Social Affairs, Division for Sustainable Development. http://www.un.org/esa/sustdev/

United States Government Accountability Office (2004). Environmental Indicators: Better Coordination Is Needed to Develop Environmental Indicator Sets That Inform Decisions. GAO, United States Government Accountability Office, Accountability Integrity Reliability, Highlights of GAO-05–52, a report to congressional requesters, November 2004. http://www.sdi.gov/lpBin22/lpext.dll/Folder4/Infobase1/1?fn=main-j.htm&f=templates&2.0

US Environmental Protection Agency (2003). America's Children: Key National Indicators of Well-Being, America's Children and the Environment: Measures of Contaminants, Body Burdens, and Illnesses, Second Edition. US Environmental Protection Agency, Office of Children's Health Protection, National Center for Environmental Economics, EPA 240-R-03–001. http://www.sdi.gov/lpBin22/lpext.dll/Folder4/Infobase2/5?fn=main-j.htm&f=templates&2.0

World Bank (2000). World Development Indicators 2000. The World Bank, Washington, DC. http://www.worldbank.org/data/wdi2000/index.htm

World Resource Institute. Environmental Risks to Human Health: New indicators. The indicators rank countries according to potential environmental threats to human health. http://www.wri.org/publication/content/8349

References

Beach, D. (2004). Coastal Sprawl: The effects of Urban Design on Aquatic Ecosystems in the United States. Pew Oceans Commission.

Blaikie, P., T. Cannon, I. Davis, and B. Wisner (1994). *At Risk. Natural Hazards, People's Vulnerability and Disasters.* Routledge.

Bolin, R. (1986). Disaster Impact and Recovery: A comparison of black and white victims. International Journal of Mass Emergencies and Disasters 4(1): 35–41.

Bolin, R. and L. Stanford (1991). Shelter, housing and recovery: a comparison of U.S. disasters. *Disasters* 15:24–34.

Booth, D. (1991). Urbanization and the natural drainage system—impacts, solutions, and prognoses. *Northwest Environmental Journal* 7(1):93–118.

Booth, D. and L. Reinelt (1993). Consequences of urbanization on aquatic systems: measured effects, degradation thresholds, and corrective strategies. In *Proceedings of Watershed '93, A National Conference on Watershed Management.* Alexandria, Virginia.

Boyd, K. and H. Stradther (2005). Assessing the Vulnerabiity of the Mississippi Gulf Coast to Coastal Storms Using an On-Line Coastal Risk Atlas. National Coastal Data Development Center: Stenner, WI.

Bruneau, M., S. E. Chang, R. T. Eguchi, G. C. Lee, T. D. O'Rourke, A. M. Reinhorn, M. Shinozuka, K. Tierney, W. A. Wallace, and D. von Winterfeldt (2003). A framework to quantitatively assess and enhance the seismic resilience of communities. *Earthquake Spectra* 19(4):733–752.

Burby, R. J. (1998). *Cooperating With Nature.* Joseph Henry Press, Washington, DC.

Carter, M., P. Little, T. Mogues, and W. Negato (2006). Shocks, sensitivity and resilience: Tracking the economic impacts of environmental disaster on assets in Ethiopia and Honduras. International Food Policy Research Institute Discussion Paper 32.

Comfort, L., B. Wisner, S. Cutter, R. Pulwarty, K. Hewitt, A. Oliver-Smith, J. Wiener, M. Fordham, W. Peacock, and F. Krimgold (1999). Reframing disaster policy: the global evolution of vulnerable communities. *Environmental Hazards* 1:39–44.

Crowards, T. (1999). An Economic Vulnerability Index for Developing Countries, with Special Reference to the Caribbean. Barbados, Caribbean Development Bank.

Cutter, S. L. (1996). Vulnerability to environmental hazards. *Progress in Human Geography* 20(4):529–539.

Cutter, S. L., B. J. Boruff, and W. L. Shirley (2003). Social vulnerability to environmental hazards. *Social Science Quarterly* 84(2):242–261.

Cutter, S. J. Mitchell, and M. Scott (1997). Handbook for conducting a GIS-Based Hazards Assessment at the County Level. Hazards Research Lab, Department of Geography, University of South Carolina.

Cutter, S. L., J. T. Mitchell, and M. S. Scott (2000). Revealing the vulnerability of people and places: a case study of Georgetown County, South Carolina. *Annals of the Association of American Geographers* 90:4 713–737.

Galli, J. 1991. Thermal impacts associated with urbanization and stormwater management best management practices. Metropolitan Washington Council of Governments, Maryland Department of Environment, Washington, DC.

Heinz III Center for Science, Economics and the Environment (2000). *The Hidden Costs of Coastal Hazards; Implications for Risk Assessment and Mitigation.* Islands Press: Washington, DC.

Hewitt, K. (1997). *Regions of Risk: a Geographical Introduction to Hazards.* Addison Wesley, Longman, Harlow, Essex.

Hoffman, S. M. (2003). The Hidden Victims of Disaster. Hazard workshop presentation, Boulder, CO.

Hossain, S. M. N. (2001). Assessing Human Vulnerability due to Environmental Change: Concepts and Assessment Methodologies. Master of Science Degree thesis, Royal Institute of Technology. KTH, Stockholm, Sweden.

Howell, S. E. (2005). *Citizen Hurricane Evacuation Behavior in Southeastern Louisiana: A Twelve Parish Survey.* Survey Research Center, University of New Orleans, New Orleans, LA.

Kasperson, R. E., O. Renn, P. Slovic, H. S. Brown, J. Emel, R. Goble, J. X. Kasperson, and S. Ratick (1988). The social amplification of risk: a conceptual framework. *Risk Analysis* 8:177–187.

Mileti, D. (1999). Disasters by Design: A Reassessment of Natural Hazards in the U.S. Joseph Henry Press: Washington, DC.

Morrow, B. H. (1999). Identifying and mapping community vulnerability. *Disasters: The Journal of Disaster Studies, Policy and Management* 23(1):1–18.

Munich Re (2002). *Topics 2002.* Report of the Geo-science Research Group, Munich Reinsurance Company, Munich.

Nakagawa, Y. and R. Shaw (2004). Social capital: a missing link to disaster recovery. *International Journal of Mass Emergencies and Disasters* 22(1):5–34.

National Research Council (2000). *Ecological Indicators for the Nation (2000),* Commission on Geosciences, Environment and Resources, National Academic Press, Washington DC. http://www.nap.edu/books/0309068452/html/.

O'Brien, P. W. and D. S. Mileti (1992). Citizen participation in emergency response following the Loma Prieta Earthquake. *International Journal of Mass Emergencies and Disasters* 10(1):71–89.

Pantin, D. (1997). *Alternative Ecological Vulnerability Indicators for Developing Countries with Special Reference to SIDS.* Report prepared for the Expert Group on Vulnerability Index. UN(DESA), 17–19 December 1997.

Parkins, J., J. Varghese, and R. Stedman (2004). Identifying indicators of community sustainability in the Robson Valley, British Columbia. *BC Journal of Ecosystems and Management* 4(2).

Pulido, L. (2000). Rethinking environmental racism: white privilege and urban development in Southern California. *Annals of the Association of American Geographers* 90:12–40.

Peacock, W. G., B. H. Morrow, and H. Gladwin, eds. (2000). *Hurricane Andrew and the Reshaping of Miami: Ethnicity, Gender, and the Socio-Political Ecology of Disasters.* Florida International University, International Hurricane Center. Miami, FL.

Richmond, B., C. Fletcher, E. Grossman, and A. Gibbs (2001). Islands at risk: Coastal hazard assessment and mapping in the Hawaiian Islands. Environmental Geosciences, Blackwell Synergy, 8:21–37.

Schueler, T. and H. K. Holland (2000). *The Practice of Watershed Protection.* Center for Watershed Protection, Ellicott City, Maryland.

SOPAC (2000). Environmental vulnerability index: development and provisional indices and profiles for Fiji, Samoa,Tuvalu and Vanuatu, EVI phase II report. South Pacific Applied Geoscience Commission.

Twigg, J. and M. Bhatt (1998). *Understanding Vulnerability: South Asian Perspectives.* Intermediate Technology Publications, on behalf of Duryog Nivaran

UNDP (2000). *Human Development Report 2000.* Oxford University Press, New York.

UNDP (2005). Reducing Disaster Risk. A Challenge for Development. United Nations Development Programme, Bureau for Crisis and Recovery.

USAID (1999). *Introduction to Current Vulnerability Guidelines,* FEWS Current Vulnerability Assessment Guidance Manual.

Walker, L., D. Zarin, N. Fletcher, R. Myster, and A. Johnson (1996). Ecosystem Development and Plant Succession on Landslides in the Caribbean. *Biotropica* 28(4a): 566–576.

Wallis, A. D., E. Aguelles, D. Lampe, and M. Meehan (2001). Imaging the region: south Florida via indicators and public opinions. Florida Atlantic University/ Florida International University Joint Center for Urban and Environmental Problems. Miami, FL. http://www.jc.fau.edu/3publications/imagingtheregion.pdf.

Williams, L. and L. Kaputska (2000). Ecosystem vulnerability: a complex interface with technical components. *Environmental Toxicology and Chemistry* 19(4):1055–1058.

World Bank (1999). Environment matters, Annual Review 1999, The World Bank, Washington, DC.

World Economic Forum (2000). *Pilot Environmental Sustainability Index.* An initiative of the Global Leaders for Tomorrow Environment Task Force, Annual meeting 2000, Davos, Switzerland, in collaboration with Yale University and Columbia University. http://www.weforum.org/pdf/glt/glt_esi_2000.pdf.

Yohe, G. and R. S. J. Tol (2001). Indicators for Social and Economic Coping Capacity—Moving Toward a Working Definition of Adaptive Capacity. Global Environmental Change 12(1): 25–40.

Zandi, M., S. Cochrane, R. Ksiazkiewicz, and R. Sweet (2006). Restarting the economy. In *Rebuilding Urban Places after Disaster: Lessons from Hurricane Katrina*, ed. E. Birch and S. Wachter. University of Pennsylvania Press, Philadelphia.

Chapter 7

Risk Communication

Objectives

The study of this chapter will enable you to:

1. Define risk communication and the communication process.
2. Examine communication barriers in discussing risk with the public or other stakeholders.
3. Examine the target audience in the risk communication process.
4. Discuss tools for risk communication including maps, figures, and community engagement.
5. Explore strategies for managing risk communication including community engagement, ethics, and decision making and legal issues.
6. Explore how organizations learn through risk communication.

Key Terms

Adaptive behavior
Credibility
Dialogue
Persuasive communication
Precautionary actions
Public information
Risk

Risk communication
Social amplification
Trust

Issue

Doubt, skepticism, and uncertainty have become part of individual views of our capacity to deal with disasters. The response to Hurricane Katrina in 2005 raised many questions about our risks associated with natural hazards. People have doubts about government policies and processes, as well as the priorities of both businesses and community-oriented agencies. As agency representatives, we can no longer just say that we know what is best. We must be sensitive to how we communicate about risk and how our risk communication strategies may be received by others. Understanding how individuals perceive risk to hazards and how the communication of risk impacts individual and organization actions is essential in reducing vulnerability to hazards. People no longer see that hazards and our vulnerability are associated with chance and factors outside human control. People appreciate that our systems are not perfect, have limitations, and may be vulnerable to human and organizational failures. We can enhance the public's understanding of hazards and risks and improve organizational adaptive measures by increasing our communication of risks. What can be done to strengthen our individual, community, and organizational resilience?

Introduction

Hazards identification and risk analysis provide a basis for profiling hazards that might impact us as well as our assets and vulnerabilities. They clarify when and where a disaster might occur and the impacts that it could have. Information from the assessment process can be used in many ways to help us adapt to our risks, including short-term hazard warnings or in the development of long-term mitigation strategies to reduce adverse consequences from disasters. A jurisdiction might initiate communication strategies to help the community know how vulnerability might be reduced through mitigation as well as what sheltering or evacuation protective actions should be taken. Many decisions associated with hazards are made on an individual, family, community, regional, and national basis. Hazards analysis does not conclude with the risk analysis, but is used to reduce vulnerability and strengthen individual, organizational, and community resilience to natural hazards. Understanding the role of risk communication in individual and organizational decision making is critical in establishing and sustaining resilient communities. Our goal is to enhance our decision-making capacity through conscious communication strategies at all levels.

Risk Communication

Risk communication involves the process of sharing information about hazards, risks, vulnerabilities, assets, and adaptive mechanisms within organizations or with the public. The process is intentional and goal directed, including sharing information about a hazard or identifying appropriate strategies to reduce vulnerability to a specific hazard. We see that risk communication involves more than just talking about the hazards, but is a process that provides a framework for enhancing our capacity to understand hazards and foster constructive adaptive strategies at the individual, community, or organizational levels to foster sustainability and resilience.

Individuals are concerned about their own safety and security and have the capacity to protect their welfare; public, private, and nonprofit agencies have the capacity to build a culture of trust and credibility to ensure that their expertise is used to support sound decision making. But failures occur at all levels, and we acknowledge that our organizations are not perfect. The key is to realize that sound decision-making is an intentional action by individuals and organizations. The risk communication process has a critical role in supporting sound decision making and the adoption of strategies to cope and deal with hazards.

> Hundreds of miles of levees were constructed to defend metropolitan New Orleans against storm events. These levees were not designed to protect New Orleans from a category 4 or 5 monster hurricane, and all of the key players knew this. The original specifications of the levees offered protection that was limited to withstanding the forces of a moderate hurricane. Once constructed, the levees were turned over to local control, leaving the USACE to make detailed plans to drain New Orleans should it be flooded (U.S. House of Representatives 2006).
> The local sponsors—a patchwork quilt of levee and water and sewer boards—were responsible only for their own piece of levee. It seems no federal, state, or local entity watched over the integrity of the whole system, which might have mitigated to some degree the effects of the hurricane. When Hurricane Katrina came, some of the levees breached—as many had predicted they would—and most of New Orleans flooded to create untold misery.
>
> *A Failure of Initiative*
> *Final Report of the Select Bipartisan Committee*
> *U.S. House of Representatives, 109th Congress (2006)*

Risk communication is more than just talking with people; it is an intentional process to gather information in order to further explain the nature and extent of hazards and disasters as well as to provide input into the decision-making process. Risk communication can thus be viewed in a broad context of hazards risk management, which is helping organizations and communities, deal with risks and reduce vulnerability to hazards and disasters.

The Risk Communication Process

McGuire (1969) provides us with a lasting approach to understanding the communication processes that are persuasive in nature and built on who says what, by a medium, to whom, and with what desired intensions. Figure 7.1 which was adapted from McGuire's approach to interpersonal communication, provides a diagram of this communication process. The key is to appreciate the source of the communication, including its credibility, trust, and authority. The nature of the message itself involves both the hard and soft characteristics, including the style, words, pace, and complexity, as well as the scientific or technical nature of the content. The medium includes how we send the message, which might involve written or oral communication through the Internet, radio or television, video conference, phone, or in person. For the person that we are communicating with, the receiver may be old or young, educated, of a different culture or ethnic background, speak a different language, and may have an interest in the subject of our dialogue. Finally, our intent in the communication process may simply be to just inform, obtain compliance with some official order, reach agreement with some future action, raise a question for discussion and exploration, or simply just form the basis for an ongoing dialogue.

The context of our risk communication may involve diseases, natural hazards, such as floods, earthquakes, fires, or drought, and may target employees, citizens, legislators or business representatives. Our communication message may involve short-term warnings of hazards or long-term awareness initiatives or efforts to raise support for changes in codes or hazard-mitigation programs.

Critical Thinking: Risk communication is person-centered in a social, cultural, and environmental context. We must acknowledge that, when we are talking about hazards, disasters, and risk, we are dealing with complex issues that affect our way

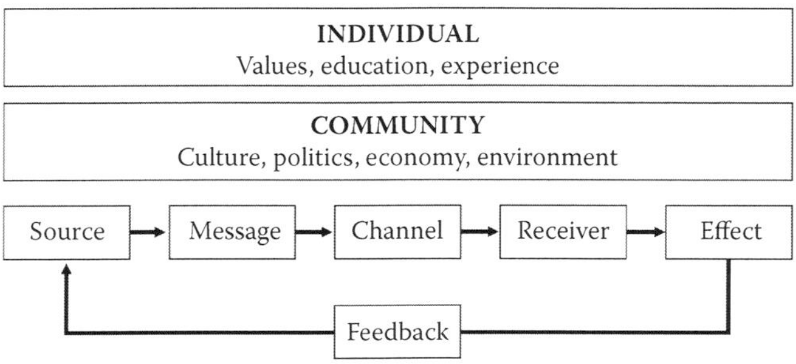

Figure 7.1 The classical persuasion model.

of life and our community. Risk communication may naturally involve conflicts between parties, and we must acknowledge how we intend on dealing with differences in our communities and within our own organizations.

As one views risk communication within a risk management context, we could characterize this communication process as persuasive in nature, since its intent is to bring about some desired action. However, this view of risk communication places great emphasis on one-way communication that results in planned outcomes. Risk communication in a risk management context changes the desired outcomes from just one-way communication to an open exchange of information and mutual understanding of complex issues. The process, in this way, becomes more of a two-way exchange of information that can lead to further clarification of issues, identification of possible alternatives to reducing the impacts of a hazard, or strategies that individuals, families, organizations, or communities could take to enhance resiliency.

Blaikie et al. (1994) note that disasters are more than just a natural event; they are the product of social, political, and economic factors. Hurricanes that strike coastal areas cause extensive destruction because of development practices and the desire to build, live, and vacation in coastal areas. Hurricane Katrina in 2005 demonstrated that the floods were more than just a physical event; it was a political, economic, social, and environmental crisis that was human influenced. Therefore, risk is more than an objective phenomenon, but one that includes social and emotional reactions of people. Individual perception of risk is thus a critical part of the risk communication process; we should be sensitive to individual perceptions if we want to truly communicate with others about disasters and our social, economic, and environmental exposures. The human aspects of risk, and how people interpret information concerning hazards and disasters, must be recognized as a critical part of risk communication. We must examine the social aspects of risk and ensure that it is included in the risk communication process.

Barriers in Risk Communication

How one views a hazard is influenced by one's own values and dynamics of power, conflict, and trust in organizations. Risk is thus highly subjective and can be perceived very differently by citizens, public officials and officers, and other agency personnel involved. Risk is a concept that is impacted by how we understand and cope with the dangers and uncertainties of life. Scientists may view risk in light of model outputs, data limitations, and assumptions. Nonscientists have their own decision rules that may be highly intuitive (Kraus et al. 1992; Morgan et al. 2002).

Our perception of risk is influenced by what we believe are the immediate direct impacts and their longer-term indirect impacts. One could have small-scale immediate damage but longer-term financial repercussions or lawsuits including recovery costs. The ripple effect of a disaster event could be long term and far reaching. How we perceive these possible effects will impact how we regard the adverse impacts of a disaster. The concept of social amplification of risk is demonstrated in limited direct impacts of events that trigger major indirect impacts (Slovic and Weber 2002).

Grothmann and Reusswig (2006) suggest that risk communication concerning the nature of an exposure to a potential hazard can result in either an individual precautionary action taken to reduce vulnerability to a hazard or a reactive adaptation following a disaster event. How one perceives information from a specific risk communication concerning a hazard may result in adaptive behavior that significantly reduces vulnerability. People who live in high-hazard zones often fail to act to reduce their vulnerability (Peek and Mileti 2002) but some households do take action to avoid damage in risk zones (Rogers and Prentice-Dun 1997). The difference may be that a minimum level of threat is perceived before a preventive action is taken (Schwarzer 1992). The key is that communities have an opportunity to influence individual and household behavior by initiating and sustaining efforts to inform citizens of the value of adaptive behavior in light of local hazards.

Critical Thinking: So what do we believe and who do we trust? How do we decide who to believe? Slovic (1993) notes that people respond to the hazards they perceive, and if these perceptions are faulty, then their actions will likely be misdirected. If we use statistics to make our case, and people do not understand their meaning, then the likely outcome will be distrust, conflict, and ineffective actions.

The Independent Investigation Team examining the failure of levees in New Orleans stressed that we should define risk within an intergovernmental framework with a focus on protecting citizens and that citizens should be included in the process of examining risk.

> Authorities for catastrophic risk management should ensure that those vulnerable have sufficient and timely information regarding their condition and a reciprocal ability to respond to requests for their informed consent especially regarding tradeoffs of safety for cost. The public needs to be encouraged to actively and intelligently interact with its development of local plans.

Kirkwood (1994) contends that there is often a difference between the object evaluation of risk and public perception. This gap is explained by the "experts" who suggest that the public just cannot understand complex scientific knowledge to evaluate risks. Unfortunately, the scientists believe that their examination of risk is rational, objective, and nonjudgmental and that risk must be explained based on technical grounds. To do otherwise would lead to gross oversimplification of risk. Kirkwood notes that unfortunately this view of expert opinion does not fit with the reality that two different experts who examine the same problem may conclude differing estimations of risk. He explains that the expert and the public look at risk very differently; the expert examines risk based on a precisely documented process, and the public by looking at potential injury, death, or loss. The public is making decisions based on their rules of thumb and subjective judgments for avoiding danger rather than a complex examination of data.

Fischhoff et al. (1982) observe that many experts believe that "people are so poorly informed (and uneducable) that they require paternalistic institutions to defend them," and furthermore, that they might be better off surrendering some political rights to technical experts. He further explains that some experts justify their unwillingness to discuss complex risks with the layman because they believe that information would make people anxious and that they could not use the information wisely if provided. Fischhoff stresses that people are very different and that we should avoid generalizations. Some are risk takers and other avoiders. Some are cautious where others are rash; it is just part of an individual's personality. He stresses that people's perceptions about risk may sometimes be erroneous, but they are seldom stupid or irrational; an individual citizen may have a different way of processing the possibility of harm and loss.

The expert and the lay person are different from one another in education level and knowledge at their disposal but not in the way they think. The experts' depth of knowledge comes at a cost in their breadth of their view of issues. One should see that the communication between the expert and the public be respectful and balanced. Effective hazard risk management requires cooperation of many lay people, and our ultimate goal is an informed citizenry. All of us benefit from careful, examined judgments including quantitative and qualitative elements. We must recognize our own cognitive limitations and temper our assessments of risk with a respectful eye to the public and openness to other views.

The best way of getting a good assessment of risk is from diverse and independent views. When decisions are made from limited perspectives the results often reveal many unexpected outcomes that were not considered. We need to be prepared for a wider discussion and address other points of view so as to mitigate a common mistrust of public institutions.

Cook et al. (2004) examined the discourse between experts and the public and came to the conclusion that scientists group the dialogue concerning risk into three groups including knowledgeable experts (scientists themselves), the public, and opponents (including the press). The public under this framework is categorized as uninformed (ignorant) and emotional rather than rational, with no understanding of risk. This view of the public allows the scientist to be free of having to engage with the public in dialogue that would be pointless, since the uninformed have nothing to contribute to a decision-making process. Opponents as a group have something at stake, are unconcerned with truth, and have nothing to gain by a dialogue. The opposition is thus discredited by scientists as not genuine.

Critical Thinking: How one frames a position is critical in any dialogue. Scientists may view a situation from empirical objectivity and consider this the only legitimate perspective. On the other hand, many issues may be framed in other ways such as morally, economically, socially, politically, aesthetically, or even scientifically. Is it justifiable? What does it cost? Who benefits? Who controls it? Does it make things more pleasing? Is it safe? Many nonscientists see many of these perspectives as very legitimate ones.

Target Audience of Risk Communication

There are many groups that may be engaged in the risk communication process from citizens prior to a disaster event or victims following a disaster. Other groups could include different public agencies, nonprofit groups, businesses, first responders, or volunteers. A strategy in communicating with these groups may change given that our goals might shift depending on who is targeted by our communication. Clarifying the audience whom we have engaged in communication concerning risk is critical in determining the content and the process of communication. Some to whom we communicate are involved in ongoing emergency preparedness, while others do not engage in emergency response. The role of the audience in the emergency management process will impact our strategy in engaging them in an ongoing process to understand and deal with risks.

- General public—the largest audience, of which there are many subgroups, such as the elderly, the disabled, minority, low income, youth, etc., and all are potential customers.
- Disaster victims—those individuals impacted by a specific disaster event.
- Business community—often ignored by emergency managers, but critical to disaster recovery, preparedness, and mitigation activities.
- Media—an audience and a partner critical to effectively communicating with the public.
- Elected officials—governors, mayors, county executives, state legislators, and members of Congress.
- Community officials—city/county managers, public works, department heads.
- First responders—police, fire, and emergency medical.
- Volunteer groups—American Red Cross, Salvation Army, the Mennonites, etc., who are critical to first response to an event.

Critical Thinking: Identify groups in your community who might want to discuss hazards and risk. Determine how they might be engaged in the hazard risk communication process. How do their roles differ? How might we engage and communicate with these different groups in helping them to understand risks and adopting appropriate action strategies in minimizing or avoiding the adverse impacts of a disaster?

Risk Communication Tools

Risk communication associated with natural and human-caused hazards is a challenging process when one attempts to explain the complex scientific elements of hazards and disasters, especially for events that have a low probability. We often look for tools and aids to help explain complex phenomena such as graphic depictions of risk or hazard maps showing areas that could be impacted in a disaster.

Communicating Risks with Maps

Hazard maps are one of our best tools to help communicate the nature and extent of risks associated with natural and human-caused disasters. The Flood Insurance Program has utilized flood insurance rate maps as a means of communicating risks associated with flooding hazards. These maps provide detailed assessments of risks within a community and clearly show areas that are vulnerable to 100-year flooding events and base construction elevations for new construction or changes in existing structures in flood plain areas.

Hazard maps come in all shapes and sizes. We use these maps to help us describe the nature and characteristics of a specific hazard (wind speed, size of storm, intensity, and related hazards) by a specific location. We use maps to explore the characteristics of the local population and their vulnerability.

A map can be an excellent tool to support our communication of risks and should include a title, a mapped area, a legend, and any credits and should provide a perspective on direction, symbols, and a scale. The map title should be short and concise. It should precisely say what is displayed in the map. The map title is usually placed above the mapped area. It is better to use a main title and a subtitle instead of one long main title. The map title should have the largest type size of any text on the map. It can be all in uppercase or in upper- and lowercase letters.

The mapped area should show a graphic representation of the cultural and physical environment and contain graphic information about a hazard or our vulnerabilities. What we represent in the map provides the content for communicating what we want someone to understand.

The map should include sources for information, the map producer, publishing date, data collection methods, information about the map projection, and other explanatory notes, etc. This information is also referred to as metadata. The legend explains all graphic representations from the mapped area. Symbols in the legend should look exactly as they appear in the mapped area (same size, color, etc.). We also include a symbol that provides direction; maps are usually oriented with north being up. As part of the content of the map, we also use symbols to represent:

- Point-like features: nuclear power plant, location of a tornado touchdown, Superfund site, etc.
- Linear features: highway, canal, hurricane track, etc.
- Areal features: wildfire, flood, landslide, etc.

Finally, we include a distance scale on the map to show the relationship between distance on the map and distance on the ground. This relationship is usually expressed in the form of a ratio relating one unit on the map (numerator) to many units on the ground (denominator). The smaller the denominator, the larger the map scale. A larger-scale map covers a smaller area, which is shown with more detail. In addition, a larger-scale map shows features from the physical and cultural environment that are less generalized. On the other hand, the larger the denominator, the smaller the map scale, the larger the area that is shown on the map, the less detail can be shown, and the more features from the physical and cultural environment are generalized.

Figure 7.2 provides an example of a map that was prepared to communicate risk to local residents, business owners, and local officials in coastal Louisiana following Hurricane Rita in 2005. The map was displayed in a local library, and meetings were held with small groups so as to facilitate communication about risks in their community. Planning for the event and what would be displayed in the map was done with local emergency management officials and representatives of Louisiana State University hazards research lab. The map provided an exceptional visualization of the coastal environment including landmarks, land elevation, political boundaries, and risk zones. Viewers of the map could easily find property of interest, geographic features that might influence their level of risk, and hurricane surge zones.

It is also an illustration of a type of map generally referred to as a thematic map, which consists of a geographic base map and various thematic overlays. This type of map is ideal for communicating relationships between hazards and known features such as roads, public buildings or parks, or water features. The map of St.

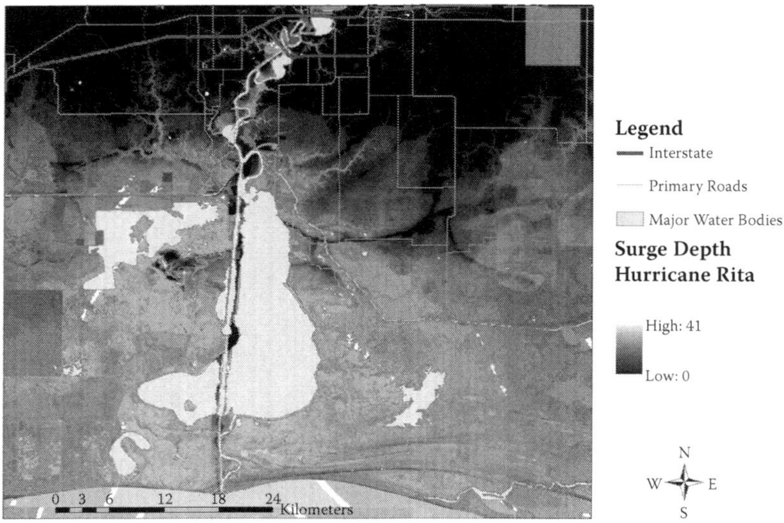

Figure 7.2 Communicating risk through thematic maps.

Mary Parish is also a qualitative thematic map, since it shows where something is located.

Reference maps such as the one shown in Figure 7.2 customarily display both natural and man-made objects from the geographical environment. The emphasis is on location, and the purpose is to show a variety of features such as coastal elevations in this map (Robinson and Petchenik 1976). A common reference map is distributed by the U.S. Geological Survey and known as digital quadrangle maps at different scales. Note the Web sites at the conclusion of this chapter for links to USGS maps. Figure 7.2 also represents a shaded relief map, since it represents physiographic features of special interest such as, in this map, differences in elevation.

The flood insurance rate map shown in Figure 7.3 is also a thematic map. Its purpose is to provide information on flood zones in an area and show base elevations where possible along a water feature. Transportation features are shown to provide a prospective on the area covered by the map. The key to all thematic maps is that they illustrate characteristics of a geographical distribution.

We use quantitative thematic maps to show how much of something is found at a location. Census Bureau data are often shown in quantitative thematic maps for a community showing population for a census block, block group, track or county boundary. A bivariable quantitative disaster map could display a dispersion plume model from a chemical incident and nighttime population density (Figure 7.4) for the same area.

Maps are able to convey information concerning hazards that may not be communicated through words. One is able to see that their community is vulnerable to a very high storm surge or heavy sustained winds that would cause extensive property damage. The map along with added discussion of the nature of a natural hazard brings the nature and scope of a specific risk to the viewer.

Figures

The use of figures to convey information is often used to explain hazards and our exposure to them. FEMA help citizens to appreciate risks by clarifying many different situations when examining the probability and consequences that hazards present. Figure 7.5 is used by FEMA to acknowledge the many situations that exist when we examine the advantages in the purchase of flood insurance under the National Flood Insurance Program (NFIP). The chart shows options that risk presents; the lighter shades have much lower risk and the darker shades higher levels of risk. If we owned property in a low-lying area near a water feature that floods often, but we have no structures on this property, then we have high probability (10) but low consequence (1). There would be no need for insurance. However, if we have a structure on the property and as the consequence of flooding increases, then flood insurance would be beneficial.

The figure helps us to judge risk by examining when a risk value of 12 for someone with a higher chance of flooding (high probability but a low consequence) has

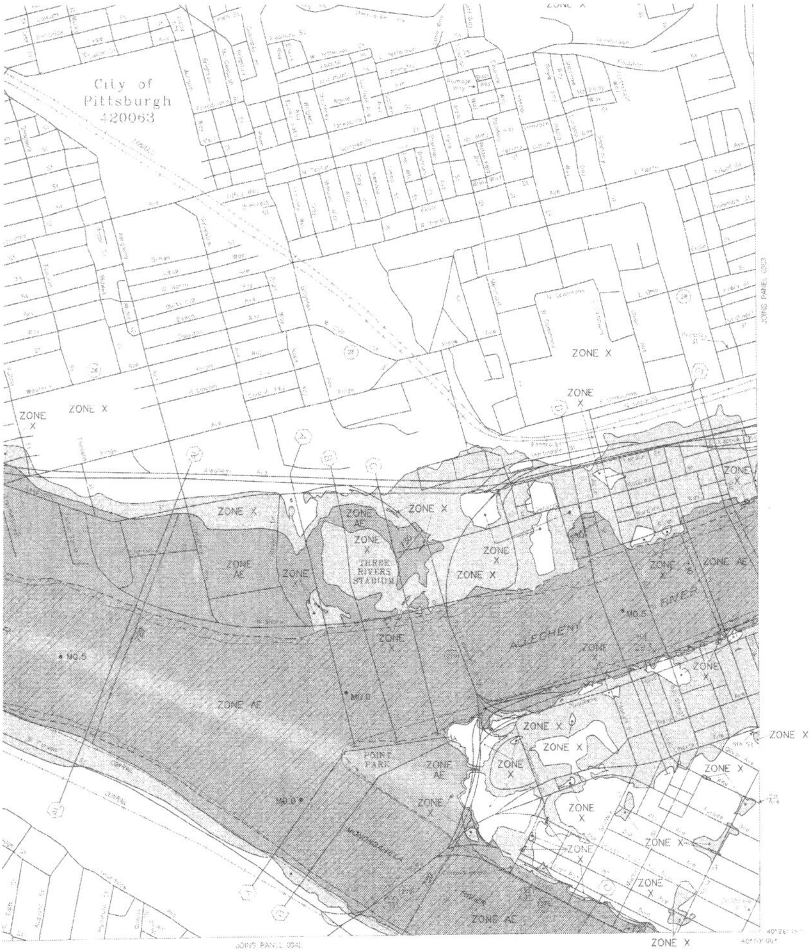

Figure 7.3 FEMA Flood Insurance Rate Map – Flood Zone

the same risk as someone with lower probability but higher consequences. We can see that if we purchase flood insurance where the risk value is 40, then through consequence management we reduce the adverse impact from flooding. The final example from Figure 7.5 shows that, when work on a levee makes it higher but a homeowner drops their flood insurance, then probability is lowered but consequence is increased by dropping the flood insurance.

Figures can facilitate communication by providing a basis for exploring different situations that a business owner, someone with property, or a renter may have about their situation. The chart thus becomes a problem-solving tool that enhances understanding of complex information.

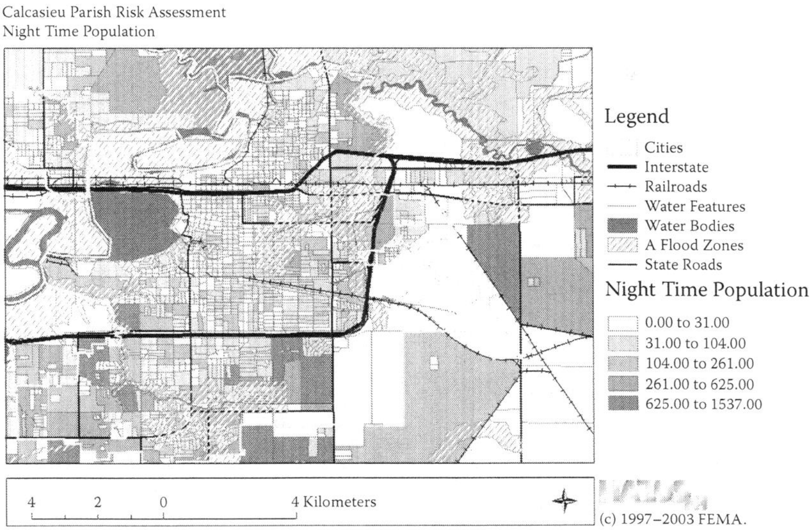

Figure 7.4 (See color insert following page 142.) Nighttime population with flood zones.

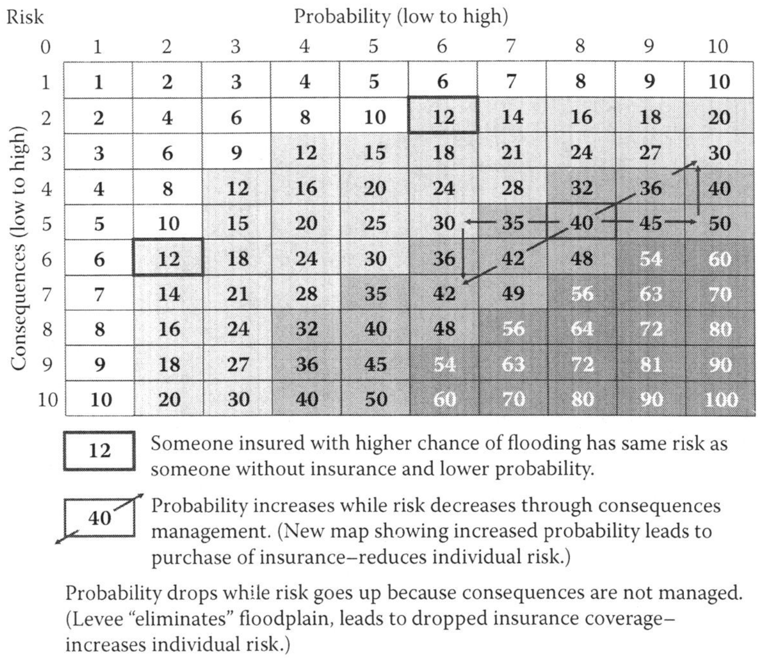

Figure 7.5 Risk probability consequence chart. Graphic design by Mary Lee Eggart.

Community Engagement

A key element of public risk communication is community engagement. Too often we see public initiatives to "sell" or "persuade" the public in something that is good for them, such as the purchase of flood insurance, adopting a family evacuation plan, securing home water heaters from earthquake hazards, or just stocking up on essentials such as water and nonperishable foods. The efforts might include the production of informative brochures, coloring books for children, incentives from utility companies to secure home gas water heaters, or newspaper flyers advocating a specific emergency preparedness suggestion or hazard mitigation initiative. In most cases, these programs to sell the public have marginal effects, and we wonder why we cannot get people to do what is smart. An alternative approach is to engage the community at various levels in a dialogue to help inform, sell, educate, or persuade families to adopt hazard mitigation efforts. We have learned in many community initiatives that involvement and engagement is the key and that including the public in the process can not only inform citizens but may also shape the form of emergency preparedness, hazard mitigation, and emergency response activities.

Community planners have known that citizens who have a stake in the neighborhood will work side by side with public employees and officials to solve problems and help identify solutions to complex problems if given the opportunity. This engagement, however, involves a two-way dialogue, where issues are identified and potential solutions examined. The result is often surprising, for a wide group of community stakeholders surface that may not have been included in the past and may be essential to implementation of a community-wide plan.

Critical Thinking: We understand that the support of the community is critical in any implementation of a public policy. Why do we spend so much time and effort on the development of the policy and program but little time in engaging the community in a dialogue on the policy or program? Give an example of a community that has successfully engaged stakeholders in a dialogue, and explain why the effort was so successful.

Community stakeholders may include small business owners, financial institutions, the school district, health care providers, local community agencies and religious organizations, manufacturing, homeowners and renters alike, and a broad range of service industry representatives. Each stakeholder has an interest and a unique perspective of risks. Identifying these stakeholders and encouraging their participation can be the key in obtaining broad-based community engagement. The role then of the public sector is to facilitate and enable the stakeholders to become and sustain their involvement.

Risk Communication Myths

A myth is something that is widely believed, but in truth, is not supported by facts. Unfortunately, widely believed risk communication myths interfere with the way we deal with others (Covello 2002) and our attempts to help communicate information about hazards and risks.

1. Communicating openly and directly with people is more likely to alarm than calm people.
   Truth: The fact is that this is not the case if information about hazards and risks associated with disasters is done properly. We need to educate and inform people and not simply alert them. Do not just show an image of a flood zone, but give people the chance to express their concerns, ask questions, and receive accurate answers.
2. Many issues associated with hazards and risk are too technical and too difficult for the public to understand.
   Truth: The fact is that a technical explanation of a hazard can confuse people despite how smart the public is. Our job is to communicate the issues no matter how complex they may be, but be conscious of how we convey information and encourage feedback and welcome questions. The public may not make technical decisions, but their opinions deserve consideration by those who are making those decisions.
3. Risk communication is not my job.
   Truth: We can find many opportunities to communicate information about risks to others, and it is our public duty to look for ways to help others understand risk.
4. If we listen to the public, we may divert limited resources to concerns that are not important.
   Truth: Listening to and communicating with the public does allow for opportunities for people to have input into organizational policy decisions. But we do not set organizational agendas and priorities based solely on prevailing public concerns. Our job is to help organizations manage issues and expectations. The public's concerns cannot be ignored, but neither can they necessarily dictate policy. The better informed people are, the more likely it will be that the public's and your opinions on priorities are aligned. Providing for opportunities for public input into policies may be a positive factor in decision making rather than a waste of time as is suggested by this myth.

Fischhoff et al. (1982) believes that most individuals overestimate their capacity to deal with disaster or any emergency. This is illustrated by the widely held belief by many that, "they are better than average drivers, more likely than average to live past 80, less likely than average to be injured by tools they operate and so on" (Svenson 1979). Although such beliefs are obviously unrealistic, people still base their perceptions of risk on their own experience and the fact that risks look very small. People still speed, run red lights, or tailgait without mishaps. Our personal experience teaches us that we are safe despite our unsafe driving practices. Our experience is shaped by daily news shows leading us to believe that when accidents happen, they happen to others. Unfortunately, we see that people do not see beyond their own perceptions, information, and experience.

Another common myth held by some scientists is that the public is ignorant of science and that they have no understanding of risk. It is perceived by many scientists that the public demands that they be exposed to "no risk." This is an assertion that is not supported by the evidence. Studies show that most people understand very well that nothing is risk free, and they are able to "live with" uncertainty and the lack of control that it entails (Wynne 2002).

Cook et al. (2004) suggest that, despite these perceptions by many scientists, the public understands that life is not risk free and that risk is present in our daily lives. As individuals representing agencies, we use risk communication to help the public to see that something is not risk free or safe, but that we are reasonably safe. A few mistakes occur, and accidents do happen despite our best efforts to prevent them.

We might state that our hazard models show that there is a very low possibility of a disaster, however, cannot say that we are without risk. We can only express that our models have not shown anything that we need to be worried about. Cook et al. (2004) see that the public are concerned about these kinds of statements. Unfortunately, many scientists believe that the public cannot understand risk, and that, if a disaster is to happen, it will not impact them.

Covello (2002) stresses that we must earn trust and build credibility with the public. We need to establish constructive communication that is based on our audiences perceiving us as trustworthy and believable. This is built on how they perceive us, including caring, antagonistic, or competent, if we are seen as honest and open, and our perceived commitment to common goals. Covello (2002) suggests that we:

Look at others as partners and work with them to inform, dispel misinformation, and to every degree possible, allay fears and concerns.

Examine their concerns. Statistics and probabilities do not necessarily answer all questions. Be sensitive to their fears and worries on a human level.

Be honest and open. Once lost, trust and credibility are almost impossible to regain.

Work with other credible sources, build alliances, and establish mutual interests.

Recognize that conflicts and disagreements among organizations create confusion and breed distrust.

Address the needs of the media. The media's role is to inform the public, which will be done with or without your assistance.

Critical Thinking: Conflict is a normal part of dealing with hazards and risks. We should expect it and understand how to cope with it in a positive manner. Conflict should not be viewed in a negative manner but as an opportunity to reveal the complex nature of the issues associated with disasters and our communities. Why is conflict such an ongoing part of our lives?

Covello (2002) provides some insights into overcoming conflict and how to deal with potential distractions using positive approaches to coping with hazards. He identifies common problems and then provides us with a suggestion for dealing with conflicts (Figure 7.6).

Managing Risks

We have stressed in this book that risks may be viewed from two perspectives including the likelihood of something happening and the consequences if it did happen (Beer and Ziolkowski 1995). We have explained that the concept of hazards evolved from the notion of games of chance. This view stresses that risks are a phenomena outside individual control and are associated with involuntariness and unanticipated consequences (Ingles 1990). Ingles goes further to stress that risk is also associated with voluntariness through individual adventurousness or individual decisions to take a chance. Risk may be managed only if we appreciate the nature of hazards and how we might be impacted economically, politically, socially and environmentally (Young 1997). A key element for anyone to cope with a disaster is the information that is provided by public, private, and nonprofit organizations through risk communication and education. What we do with our understanding of hazards and risks impacts our capacity to deal with disasters on an individual, organizational, or community level. We therefore cannot manage risks that we do not understand.

An appreciation of our social, economic, and environmental vulnerability is critical in selecting and implementing adaptive strategies to reduce the adverse impacts of disasters. Vulnerability involves the capacity to withstand a hazard and, for individuals, families, and communities, the capacity to cope with hazards. Hazards risk management involves strategies to strengthen community building codes and their enforcement, adopt planning guidelines, purchase a flood insurance policy, invest in elevating an existing or new structure above anticipated flood levels, relocate critical storage areas, or develop operational plans if transportation routes are closed. Through risk management, we see that we can adopt and implement specific actions that can reduce vulnerability and enhance our resilience to disasters. When disaster strikes, we are caught up in reacting to the damage rather than looking for ways to mitigate or prevent the damage from occurring.

Discussions of risk can be very abstract.	We should frame our points by using examples, stories, and analogies.
Conflicts that are associated with hazards and risks can result in attacks.	Respond to issues, not to people. Strive not to further debates.
We may convey how upset we are through our non-verbal cues and messages.	We need to remain calm, attentive, and polite. Adopt a relaxed, neutral physical stance. As long as we don't express our anger, we should be able to deal with the conflict.
Defensiveness never pays off in conflict and when we blame others it just shows that we are attempting to defend ourselves.	The best answer is to just accept your share of responsibility for a problem.
Defensiveness comes in many forms and blaming others is just one example, a second is attempting to deal with conflict by using costs as a justification for our position.	Focus on the benefits to be derived. If costs are an issue, voice respect for the need for responsible stewardship of public funds.
When we are pressed, we may be tempted to provide commitments that we just cannot guarantee.	It is better to express your intentions in terms of what you want to accomplish. Stick to the facts of what has, is or will be done.
People representing organizations often resort to the use of jargon to communicate their message.	Avoid the jargon, define technical terms and don't use acronyms.
We assume that what we say is not part of the public record.	Avoid making a statement that is "off the record."
Statistics can be confusing.	Use visuals, hand-outs, or displays to convey your message and don't try to just explain complex information by words alone.
Technical details and data can inform but they can easily confuse people on the points that you want to stress.	Avoid trying to fully inform and educate audiences on the minutia of issues.

Figure 7.6 Conflict situations and coping strategies. Graphic design by Mary Lee Eggart.

Decision Making

We must acknowledge that there are no guarantees that our decisions and actions will reduce all damages. Unfortunately, there is always a possibility that failure can occur and that risk cannot be completely eliminated (Barnes 2002). The key is that we must build a process for enhancing our understanding of risk, and that information about hazards and their impacts can support short-term and long-term decision making that acknowledges our vulnerabilities and risks to hazards.

Organizations must adopt a position of risk management as a strategy to understand the nature of hazards. Risk communication must be part of the hazards identification process where we share our community assets, resources, and the nature of hazards that could impact us. It is also part of risk analysis where we explore our vulnerabilities through spatial analysis tools and techniques. Risk analysis thus becomes the basis for us to adopt sound decisions and systems.

Risk communication within our organizations and with the public is a critical part of our attempts to manage hazards and to deal with risk. The intent is to ensure that everyone understands hazards and what can be done to reduce our vulnerability on an individual, community, regional, or national level from warning systems to forecasting or crisis decision-making processes (Plate 2002). We thus need to combine the best science and decision support systems with an ongoing dialogue with our community.

Community Engagement in Hazards Analysis

We have stressed that risk is associated with our actions or inactions, natural events, and our use of technology. Although each disaster event is unique, each involves human decisions that either increase or decrease our vulnerabilities. Risk communication throughout the hazards analysis process facilitates our capacity to identify and implement sound strategies to manage risks.

One of the key questions is who should be involved in the hazards analysis process and in making decisions that result from this process. This question is best resolved by examining the goals of the hazards analysis and who is needed to provide you with a quality effort. In order to attain the goals, you may need to involve a variety of people from throughout the organization and even outside.

Another look at the hazards analysis process in Figure 7.7 shows that it involves risk communication and community engagement throughout the process from hazard identification, risk analysis, and the identification and adoption of coping mechanisms. If stakeholders are to be involved, it needs to begin from the start and have them help explain the unique human and cultural capital of the community, economic and critical infrastructure, and especially what elements of the ecosystem capital make the community unique.

As we move into the development of hazard profiles for the community, would community engagement strategies be the same as with the identification of community assets? Chances are that the identification and description of community hazards needs to involve skilled community planners, GIS specialists, modelers from the local university, or extension outreach specialists as well as emergency management staff at the local or state level. Are there other stakeholders who have specialized knowledge of either geologic, hydrologic, climate, or hazardous

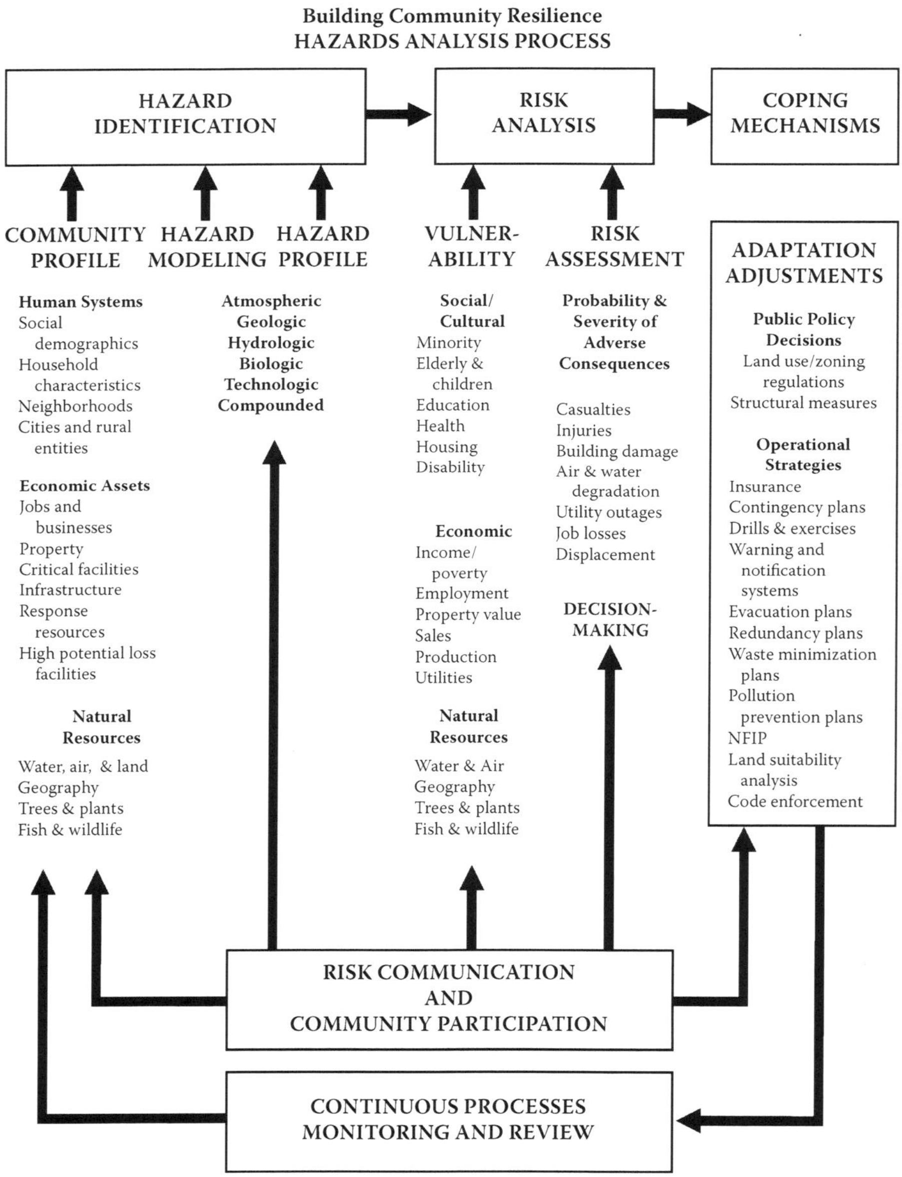

Figure 7.7 Community involvement in the hazards analysis process.

materials hazards and can be called in to help? Even the local school system might be engaged to help explain the nature and extent of a local hazard as part of a science project. This engagement can enlighten the school district on hazards that had not been fully appreciated and showcased in the local media.

One of the key elements of decision making is determining who if anyone is involved in the process. Given that we live in an open society and that there are many interested parties in the results of a hazards analysis, involving stakeholders in the process is not uncommon. The question then centers on who is involved and the process that best fits with their role in dealing with organizational or community risk from hazards. To what extent and how should residents and representatives from business and nonprofit organizations be involved in analyzing risks, defining the problem and decision making?

There are many special interests associated with any examination of risks and hazards; how should these be considered? To what extent should they be discussed in an open arena? How to include stakeholders in the hazards analysis process and the problem-solving process is essential if acceptance and understanding of the output of the analysis is a priority. Risks associated with natural hazards are multidimensional and affect stakeholders from numerous public, private, and not-for-profit agencies. The process of decision making is thus as important as the final results of the analysis.

Unfortunately, we do not anticipate the degree of conflict that may arise in attempting to find ways of reducing the adverse impacts of disasters. It is therefore crucial that we manage the conflicting forces that are stirred up by the alternatives that we are considering. There are many reasons that we encounter conflict in decision making; one involves failing to reach agreement on our goals, priorities, and essential elements of a solution. It is crucial that we take our time in examining alternatives and adequately account for the unknown nature of many disaster impacts. Many outcomes are possible.

Petak notes that conflicts in determining acceptable mitigation or risk management solutions center on the fact that hazards always involve risks, and to resolve disputes, one must develop and integrate as much information as possible about existing conditions and potential future outcomes. He observes that current approaches to problem solving place a great deal of power in the hands of the technical expert and professional administrator. He suggests that elected officials must be included and assume their responsibility for protecting public interests. He stresses that it is not possible to be risk free, for there are no zero risks to natural hazards. What makes this process difficult is that there are two sides of dealing with risks: one that is empirically verifiable and associated with facts and the other that is based on individual judgment. Acknowledging these two perspectives contributes to a more balanced discussion of risks and reaching acceptable solutions to problems presented by disasters (Petak 1985).

Stakeholder Involvement

One of the key issues in risk analysis is the process that is selected and who is involved in the situation associated with the problem and thus this problem-solving process. Technical versus democratic approaches or models to risk analysis provide two very different ways of determining the level of risk that an organization or community face. One is based on experts and scientific fact; where the other includes a more democratic and participatory model involving lay persons or the public along with the experts. If our goal is to have a risk analysis that is acted on by the public, then the latter approach is one that should be considered. Including stakeholders from the organization or the community is a small price to pay for the added benefits associated with different perspectives examining a very complex problem. An alternative may be an adversarial model that provides opportunities for the public to react, respond, or give input following a presentation by the scientific experts.

Because the process of determining risk acceptability (including mitigation spending and regulatory practices) is one that is influenced by many different interests, it is possible to include a broad range of groups to participate in defining what acceptable risk is for a local community or organization. Increased public participation in the process can broaden the discussion as to what constitutes acceptable risk. Unfortunately for organizations and communities, it is becoming more difficult today to deal with hard decisions relating to hazards and our vulnerability. The key question centers on just how safe is safe enough. Determining what is acceptable risk is difficult to resolve, because each interest group in a community or organization has their own priorities, biases, social values, and resources. More importantly, conflict over what is acceptable is often the result of relevant facts that allow us to view complex problems. The problem is further complicated because even experts often disagree on the degree of risk that we face from a hazard. Techniques to overcome conflicts associated with acceptable risk may result in common approaches to risk management. The key is to have a process in which to review the nature of risks associated with natural hazards.

Fischhoff (1982) suggests that we consider the following when attempting to deal with acceptable risk.

- Recognize the complexities of the nature of acceptable risk.
  - Acknowledge many approaches and perspectives in dealing with risk.
  - Utilize more than one approach in examining risks and what is acceptable.
  - Determine a basis for quantifying risk.
- Develop a decision-making process that matches the expectations of those involved.
  - Identify the role of government in risk.

One of the fundamental barriers in dealing with acceptable risk is that one or more of our alternatives to dealing with the problem may threaten the life, health, welfare, or property of others.

There is a real benefit in evaluative our thinking, for it drives us to look deeper in our options in addressing our problem. Developing a structure for comparing alternatives is seen in examining the costs and benefits of our choices. This cost benefit analysis may need to provide for future values and for estimating indirect costs of a solution. The process provides a constructive quantitative means in comparing one approach to another. But more will likely need to be considered, such as what our selected approach represents in terms of our values and priorities and whether this solution provides a precedent that is unwelcomed.

Engaging community stakeholders in the risk analysis process is also essential, for they have their own views of the probability of a hazard event and the significance of the consequences of a disaster. Soliciting stakeholders in the risk analysis process not only improves the quality of the analysis but also leads to a more thorough understanding of risk and potential strategies that might address them.

A key means of engaging stakeholders in the hazards analysis process is present in every community as a result of the passage of the Community Planning and Right-to-Know legislation in 1986. This statute evolved from national efforts to enhance planning for human-caused, technological hazards associated with the threat presented by hazardous chemicals. State emergency preparedness committees were created by this national legislation, and local community planning committees (LEPC) were formed. Membership on local LEPCs included representation of stakeholders from both the public, private, and nonprofit sectors. Each person on the LEPC has something to contribute and can be a link to the broader community. The membership of the LEPC usually includes emergency responders representing police, fire and emergency medical, health care, media, chemical processing businesses, and education. The LEPC provides a natural means of engaging stakeholders in the hazards analysis process for not only human-caused disasters but also for natural hazards.

Barnes (2002) states that many communities experience a common conflict over known or suspected links between chemical hazards and perceived health impacts. He sees that there is a public disbelief when public officials or business representatives state that there is minimal risk from hazardous substances.

Barnes notes that the public may have fears of large scale industrial disasters or long-term chronic exposures to hazardous substances in the community. The public is not convinced by authorities that the public faces more obvious threats to health and well-being existing in everyday life. Barnes goes on to explain that many communities perceived that there is inequity in exposure to harm, especially to low-income residents. Trust and credibility are the core issues impacting the perception of risk in the community and how strategies and policies to address risks from hazards should be constructed. He sees that there may be a technical rationale concerning a hazard along with a cultural perception of risk in the community that, unless addressed, could cause extensive conflict.

The potential for conflict is observed by Krimsky and Plough as a difference in technical and cultural views of risk (1998). For example, a technical perspective views risk through scientific methods, evidence, and explanations, while a cultural rationale views risk from a political one. Technical perspectives utilize organizational authority and professional expertise, while cultural perspectives appeal to one's wisdom or cultural traditions. Technical understanding evolves from quantitative methods, while cultural ones evolve from personal interviews or experience. One view suggests that it is the role and responsibility of the expert and regulatory agencies for decisions about and regulation of risks associated with health hazards. This is a view that the professionalized bureaucracy and the scientific community knows best. This is in contrast with a citizen's view concerning risk that is shaped by cultural influences, personal knowledge, or unarticulated views of hazards. Figure 7.5 illustrates the dimensions of the two perspectives on hazards.

Unless these differences in how one perceives risk are surfaced and articulated in an open setting and one that promotes communication and trust, conflict or polarization can develop in the community. The issues presented by Plough and Krimsky concerning health risks from chemical hazards could easily evolve from risks from flood, wildfire, earthquake, or other technical complex hazards. Without broad-based community engagement in an open dialogue concerning risk, it is easy to see that conflict will evolve between the technical elite and the community.

Our attempts to manage risks may evolve into measures to contain it as well as to open communication about risks. We find that the past can illuminate failures, their causes, and their control as lessons for engaging new issues and threats. The future commands the exercise of foresight, an imaginative process involving possible scenarios stirred by such questions of what if, or what could happen if, or might this occur under these circumstance. These questions help us to see options and the possibilities and allow us to identify alternatives to dealing with risks. Risk communication thus facilitates opening our perspectives and horizons to how we might manage and deal with the future.

Ethics and Decision Making

Ethics raises issues concerning how we are to act and under what standards we should base our decisions. In making decisions, we reflect on what we believe is right or wrong and the standards that these judgments are based upon. Our decisions and the manner in which they are made reflect what we believe is important and what groups, issues, or facts we ignore. Ethics is then the study of our standards that we adopt and that serve as a basis for what we consider right, suitable, and appropriate (Velasquez et al. 1987).

When we are considering ethics in making decisions, we take a look at how others may be impacted or how the environment (including property, animals, the air, or water) might be impacted. These concerns may be viewed at a personal level, institutional, or societal level.

Ethical considerations are not limited to legal concerns of what is right or wrong, for we look at the rights of citizens, employees, investors, as well as our institutions.

Technical Rationale		Cultural Rationale
Trust in scientific methods, evidence, and explanations.	↔	Trust in political culture and democratic process.
Appeal to authority and expertise.	↔	Appeal to folk wisdom, peer groups, and cultural tradition.
Boundaries of analysis are narrow and reductionist.	↔	Boundaries of analysis are broad and include use of analogy and historical precedent.
Risk is depersonalized focusing on measures of statistical variation and probability.	↔	Risk is personalized with emphasis on impacts on the community and family.
Concerns and issues that can't be described or clearly expressed are irrelevant.	↔	Unanticipated or unarticulated issues or concerns are relevant.

Figure 7.8 Technical versus cultural rationales on risk communication.

The key question in decision making and ethics is what standards of right and wrong we are including in making judgments. These standards may relate to obligations, benefits that result, or the degree of fairness to parties associated with the decision.

We make ethical judgments throughout the hazards analysis process, from determining what we consider when we characterize the local community and what is at stake when a hazard could be present. We judge what is suitable and appropriate when we select one methodology over another when our approach could result in some form of harm. Our standards of right and wrong may also influence the degree that we consider potential hazard impacts on various environments, social groups, or who wins or benefits from our analysis in the commercial sector. Central to ethical decision making is determining when our standards of right or wrong influence our judgments and have an impact on people, enterprises, society, and the environment.

Critical Thinking: Take a look at the potential impacts of our decisions which are a part of the hazards analysis process and determine who benefits and who might suffer some loss. Are all stakeholders treated equally, or are some benefiting from your assessment of hazards and their associated risks? Reflect on the outcomes of your hazards analysis and determine if all interests have had the opportunity for input equally. Do your recommendations and decisions reflect the common good? Given that many stakeholders in the hazards analysis process have very different positions and interests, do your recommended risk management strategies or hazard mitigation initiatives suggest that some parties will benefit at the price of others? How could your conclusions, recommendations, and decisions support the common good and be seen as impartial?

Legal Issues in Decision Making

Emergency management does not exist in a vacuum but is part of a broader context of federal and state statutes, court decisions, administrative rules, and constitutional provisions. In addition, local jurisdictions adopt ordinances and administrative procedures. Agency statutes or regulations may require that a comprehensive hazards analysis be conducted as a part of emergency planning or hazard mitigation efforts. A comprehensive hazards analysis for a community or organization may support zoning, changes in building codes, or other public policies.

In the past, public entities enjoyed immunity from civil claims for financial compensation because of the negligence of governmental agencies, officials, employees, or volunteers. This protection, known as sovereign immunity, was based on a belief that neither the king nor his representatives could do any wrong. Under this doctrine, individuals or businesses that were harmed by the actions of public entities are prohibited from recovering damages from the government, which was thus immune from suit (Reynolds 1982). Public agencies covered by sovereign immunity may pay claims, but they are not required to do so.

State laws have changed over the past forty years, making public entities at the state and local governments more accountable for their actions (Pine 2005). Most states adopted tort claims acts that provide some immunity for agencies and employees but no longer provide total protection for state and local public officials, employees, or volunteers. In the case of state or local officials participating in a hazards analysis, state law governs their potential liability, and almost all states provide statutory immunity for being involved in emergency management activities. It should be noted that this statutory immunity also covers volunteers who assist in emergency management activities such as a hazards analysis.

A citizen who wants to file a claim against a public agency or employee (including volunteers) files a tort claim against the jurisdiction sponsoring the hazards analysis. The claim would state that they suffered some harm because of the negligence of public officials or employees who participated in the hazards analysis or who made decisions based on the analysis. The claim of negligence would state that the public official or employees did not exercise an appropriate degree of care, skill, and diligence, which the reasonable or prudent person would exercise under similar circumstances. The injured party would state that they suffered property damage or personal injury as a result of decisions that were based on the hazards analysis and that those doing the hazards analysis were negligent.

This rule, as applied to governmental entities and those who act on their behalf (volunteers), must be understood in terms of the essential elements of negligence. All claims of negligence include:

1. Duty: Failing to conform to a defined standard of care owed to others and established by statute, common law (based upon judicial decisions), or a policy by the governmental entity.
2. Breach: Failing to conform to that standard of care or failure to carry out the required duty.
3. Damage: Actual loss or damage to the injured party(ies).
4. Causation: There must be a connection between the act of the governmental employee, official, or agency body and injury to a third party(ies).

All negligence cases have these elements in common, and absence of proof of any one element will defeat a finding of liability (American Law Institute 1965).

Many state emergency management acts require local jurisdictions to prepare a comprehensive emergency response plan. Given that state law imposes a duty to plan on local entities, failing to prepare a plan could leave the community open to a claim of negligence. Emergency plans that are out of date, have not been based on realistic assumptions, inflate capabilities, do not involve all operations areas and functions of governmental entities, or are not fully developed could be the basis for a claim. The court would have to determine if the local jurisdiction met their statutory duty to plan.

Critical Thinking: A local public entity that does a good hazards analysis that forms the basis of the local emergency plan is taking a positive step in satisfying the duty to plan. The court would examine the work of the public entity and judge if their actions were reasonable and prudent under the circumstances. Doing a comprehensive and complete hazards analysis is a positive action that would demonstrate an intent to meet the duty imposed by the emergency management statute. Involving trained people in the hazards analysis process, using accepted hazard models and risk analysis techniques, having capable people review the results of the hazards analysis, and testing the assumptions of the analysis would all be symbols of responsible actions. Having a realistic, effective, and current hazards analysis and plan is one of the best ways for a community to reduce its liability exposure.

Note that the intent of the statutory immunity provision in state laws is intended to encourage local officials to do their jobs. Legislatures recognize that emergency management is complex and that it involves discretionary judgments. Immunity is intended to encourage local officials to do their best and not be distracted by the threat of lawsuits. Further, public officials are asked to use their judgment and execute their jobs in a professional manner.

State law provides extensive protection to public officials and employees; however, there are limitations to the protections. The first exception is for willful misconduct, which includes actions that are intentional or completely unreasonable under the circumstances or actions that area consider gross negligence. Gross negligence is not mere thoughtlessness, inattention, or a mere mistake resulting from inexperience or confusion, but actions that are completely unreasonable under the circumstances.

The best approach in dealing with potential liability is to include trained staff in conducting the hazards analysis, get internal and external feedback on the results of the analysis, involve stakeholders, and share the results with the public. Document how you conduct your hazards analysis and note why you did what you did throughout the process.

Indemnification

Public officials and representatives of governmental agencies involved in a hazards analysis are generally entitled to protection against personal financial loss or indemnification if they are subject to a claim of negligence while working on the community hazards analysis. Almost all states recognize that the governmental unit is responsible or liable for the negligent acts or omissions of its agents or employees who are acting within the scope of their duties as public employees. Employees in this context include not only paid staff, but also volunteers who participate in the hazards analysis process. The public entity is thus vicariously liable for the actions of government officials and employees.

Critical Thinking: One final note on liability and claims of negligence should be examined. Local governments retain the services of legal counsel who has a clear understanding of the law of torts under state law. The General Counsel is a key person to know, for they know state laws, which apply to local government and its employees. Public employees and volunteers involved in a community hazards analysis should communicate with legal council and explain what they do and how they operate. Legal counsel can clarify state law and if additional steps in the hazards analysis might be taken.

Acknowledging Risk as a Part of Risk Communication

Hurricane Katrina demonstrated that many citizens and public officials alike failed to appreciate the nature and degree of risk in living in an area surrounded by man-made levees. Clearly, we have much to learn from how we have dealt with hazards such as coastal storms, floods, earthquakes, or wildfires in the past. Despite the fact that we have reduced fatalities from disasters in the United States, we have seen drastic increases in property damage and economic losses of all kinds.

Examining the past and benchmarking is one way to understand and evaluate our hazard policies and practices. Our past policies and practices have influenced how communities have been impacted by disasters. Our past beliefs, practices, policies, and how our organizational systems interfaced all form the basic elements of community functioning. The fact is that many of our losses from disasters are imbedded root causes that are part of our public policies, business practices, and social values in communities. Katrina revealed more fundamental problems in New Orleans that were part of the community educational, medical and mental health, emergency management, public safety, and public housing systems. Reversing root causes that are part of very complex systems is not a simple fix such as adapting a building, bridge, or water feature. Patterns of behavior, practice, and policy that help to create vulnerability are often pervasive throughout the society and invisible to our normal view of a community. The root causes of vulnerability can be reversed by recognizing problems in our systems and approaching them in an ongoing systematic way.

Our work needs to begin with a judicious and honest assessment of threats, followed by investments in prevention and mitigation and by construction of response systems that will be equal to a larger class of disturbances than we have previously allowed ourselves to contemplate (Leonard and Howitt 2006).

We know that technological failures virtually always occur within the context of management failures, and there is a growing body of literature that describes management implementations designed to reduce large-scale failure (e.g., Dekker 2002; Roberts and Bea 2001; Weick and Sutcliffe 2001).

The Hurricane Katrina catastrophe exposed a technological failure of inadequate defenses against a predictable, risky, and potentially lethal event. Recent studies have tended to focus primarily on death and destruction from flood waters released by collapse of the levees. Studies of cause acknowledge the extreme forces of nature, but also cite human and organizational errors. Human and organizational failures far exceed mechanical sources in the overall Katrina catastrophe (Seed et al. 2006). Katrina is viewed by many as an unforeseeable catastrophe; however, it only shows that we need to be ready for the worst that nature or evil men can throw at us.

Learning as a Part of Risk Communication

Learning from documented failures is a powerful method for reducing risks of repeated losses. Another method is to learn from close shaves. Many dangerous events fortunately culminate in only an incident rather than an accident, but the repetition of similar incidents can serve as early warnings of danger. Indeed, the logging and analysis of such events on the nation's airways partially accounts for commercial aviation's impressive safety record. A system for reporting close encounters of aircraft was installed decades ago. Anticipating the possibility that perpetrators of high-risk events might be reluctant to blow the whistle on themselves, many years ago the Federal Aviation Administration arranged for NASA to collect incident data and to sanitize it to protect the privacy of the incident reporter. NASA also screened reports to identify patterns as early warning of dangerous conditions. Similar systems are in place for reporting nuclear power plant incidents.

With the growing recognition of human factors in accidents or in failures to limit damage, a class of situations entailing uncommonly high risks but conspicuously good safety records was examined. In the Navy, for example, high risks are a part of daily operations of submarines and aircraft carriers. Yet accident rates are paradoxically low. Careful analysis of these situations showed that certain qualities of leadership and organizational culture foster integrity, a sense of responsibility among all participants, a tolerance by authority figures for dissent, and consensus on common goals of safe performance. High safety performance is associated with an institutional culture that is bred from the top of the management pyramid. The most critical element of that culture is mutual trust among all parties (e.g., Roberts 1990).

Discussion Questions

What is the link between hazards analysis and risk communication?

Why do experts and the public have such a difficult time in communicating?

Why is conflict such an ongoing element of the risk communication process?

Risk may be highly subjective and perceived very differently by citizens, public officials and officers, the press, and others. How might each view risks associated with natural hazards?

Why is it that some people take precautionary actions when faced with hazards, and others do nothing?

Why are hazard maps such a useful tool in helping to communicate the nature and extent of risks associated with natural and human-caused disasters?

The Independent Investigation Team examining the failure of levees in New Orleans stressed that citizens be involved in discussing risk. How do you get the public engaged in a discussion of risk? How should the community be involved in the risk communication process? Should it be a one-way process or a dialogue? How do you ensure that this process is a two-way dialogue?

What barriers inhibit the communication of accurate and timely data to decision makers? What inhibits getting information to the right places at the right time?

Most states provide immunity to state and local emergency management agencies for carrying out emergency management functions, such as planning. Assume that the hazard analysis is completed by well-trained, capable public agency staff members using widely accepted models and risk analysis techniques, but that errors in a hazards model were not discovered until after a disaster. Could the local government be liable?

What might be an example of "willful misconduct" by a local official or employee involved in a hazards analysis?

Applications

Search for maps with different scales on the Internet and find examples of cultural and physical features that are represented with different categories of symbols (point-like, linear, or real) at different scales.

What are some examples of qualitative and quantitative thematic (disaster) maps? For example, a qualitative thematic (disaster) map could show the different types of disaster (flood, forest fire, hurricane, etc.) for an entire country or a continent. A quantitative thematic (disaster) map could visualize the amount of damage one or more disasters caused in each state of the United States.

Web Sites

Agency for Toxic Substances and Disease Registry. ATSDR—health risk communication primer. http://www.atsdr.cdc.gov/risk/riskprimer/index.html

Centers for Disease Control and Prevention. CDC Emergency and Risk Communication. http://www.bt.cdc.gov/erc/

Center for Risk Communication Research, University of Maryland. http://www.comm.risk-center.umd.edu/

Govern Accounting Office. *Homeland Security: Communication Protocols and Risk Communication Principles Can Assist in Refining the Advisory System*. http://www.gao.gov/new.items/d04682.pdf

Health and Risk Communication Center. College of Communication Arts and Sciences, Michigan State University. http://hrcc.cas.msu.edu/

Institute for Risk Analysis and Risk Communication. http://depts.washington.edu/irarc/. Risk World. Risk communication Web sites. http://riskworld.com/websites/webfiles/ws5aa014.htm

U.S. Army Center for Health Promotion and Preventive Medicine. HRC home. Health risk communication. http://chppm-www.apgea.army.mil/risk/

U.S. Department of Health and Human Services (2002). *Communicating in a Crisis: Risk Communication Guidelines for Public Officials*. U.S. Department of Health and Human Services, Substance Abuse and Mental Health Services Administration. http://www.riskcommunication.samhsa.gov/RiskComm.pdf

U.S. Department of Health and Human Services. Risk communication. Public health communication during a pandemic flu. http://www.pandemicflu.gov/news/rcommunication.html

References

American Law Institute (1965). *Restatement of the Law: Torts,* 2nd Edition. American Law Institute Publishers, St. Paul, MN.

Barnes, P. (2002). Approaches to community safety: risk perception and social meaning. *Australian Journal of Emergency Management* 17(1):15–23.

Beer, T. and F. Ziolkowski (1995). Environmental Risk Assessment: an Australian Perspective, Report to Environment Protection Agency, AGPS, Canberra.

Blaikie, P., T. Cannon, I. Davis, and B. Wisner (1994). *At Risk: Natural Hazards, People's Vulneraility and Disasters.* Routledge, London.

Cook, G., E. Pieri, and P. T. Robbins (2004). The scientists think and the public feels: expert perceptions of the discourse of GM food. *Discourse Society* 15:433–449.

Covello, V. (2002). Myths, Principles, and Pitfalls. In *Communicating in a Crisis: Risk Communication Guidelines for Public Officials.* U.S. Department of Health and Human Services, Substance Abuse and Mental Health Services Administration, pp. 23–34.

Dekker, S. (2002) *The Field Guide to Human Error Investigations.* Ashgate, Aldershot, England.

Fischhoff, B., P. Slovic, and S. Lichtenstein (1982). Lay foibles and expert fables in judgments about risk. *American Statistician* 36(3) Part 2:240–255.

Grothmann, T. and F. Reusswig (2006). People at risk of flooding: why some residents take precautionary action while others do not. *Natural Hazards* 38:101–120.

Ingles, O. (1991). A Linguistic Approach to Hazard, Risk, and Error. In *New Perspectives on Uncertainty and Risk.* eds. J. Handmer et al. Centre for Resource and Environmental Studies, Australian National University and Australian Counter Disaster College Natural Disasters Organisation, 66–78.

Kirkwood Alan S. (1994). Why do we worry when scientists say there is no risk? *Disaster Prevention and Management.* 3(2):15–22.

Kraus, N., T. Malmfors, and P. Slovic (1992). Intuitive toxicology: expert and lay judgments of chemical risks. *Risk Analysis* 12:215–232.

Krimsky, S. and A. Plough (1988). *Environmental Hazards: Communicating Risks as a Social Process.* Auburn House, Dover, MA.

Leonard, H. B. and A. M. Howitt (2006). Katrina as Prelude; Preparing for and Responding to Future Katrina-Class Disturbances in the United States. Testimony before the U.S. Senate Homeland Security and Governmental Affairs Committee, Washington, DC.

McGuire, W. (1968). Personality and susceptibility to social influence. In E. Borgata and W. Lambert (Eds.) Handbook of Personality Theory and Research. Rand-McNally, Chicago, 1130–1187.

Morgan, M. G., B. Fischhoff, A. Bostrom, and C. J. Atman (2002). *Risk Communication: A Mental Models Approach.* Cambridge University Press, New York.

Peek, L. A. and D. S. Miletik (2002). The history and future of disaster research. In *Handbook of Environmental Psychology*, eds. R. B. Bechtel and A. Churchman. John Wiley and Sons, New York.

Petak, W. J. (1985). Emergency management: a challenge for public administration. *Public Administration Review,* 45(Special Issue):3–7.

Pine, J. C. (2005). *Tort Liability Today: A Guide for State and Local Governments* 10th Edition. Public Risk Management Association, Washington, DC.

Plate, E. J. (2002). Flood risk and flood management. *Journal of Hydrology* 267:2–11.

Reynolds, O. M., (1982). *Handbook on Local Government Law,* 2nd Edition, West Publishing Co., St. Paul, MN.

Robinson, A. and B. Petchenik (1976). The Nature of Maps: Essays toward understanding maps and mapping. University of Chicago Press: Chicago.

Roberts, K. H. (1990). Some characteristics of one type of high reliability organization. *Organization Science,* 1:160–176.

Roberts, K. H. and R. G. Bea (2001). Must accidents happen: lessons from high reliability organizations. *Academy of Management Executive,* 15:70–79.

Rogers, R. and S. Prentic-Dunn (1997). Protection Motivation Theory. In Handbook of Health Behavior Research, D. Gochman, (Ed.), Plenum: New York, 1:113–132.

Seed, R. B. et al (2006). Independent Levee Investigation Team. Organized for Success. *Investigation of the Performance of the New Orleans Flood Protection Systems in Hurricane Katrina on August 29, 2005.*

Slovic, P. (1993). Perceived Risk, Trust, and Democracy. *Risk Analysis,* 13(6):675–682.

Slovic, P. and E. Weber (2002). Perception of risk posed by extreme events. *Risk Management Strategies in an Uncertain World,* Palisades, NY.

U.S. House of Representatives (2006). *A Failure of Initiative. Final Report of the Select Bipartisan Committee.* 109th Congress.

Velasquez, M., C. Andre, T. Shanks, and M. J. Meyer (1987). What is Ethics? *Issues in Ethics* IIE Vol. 11.

Weick, K. E., and K. Sutcliffe (2001). *Managing the Unexpected,* Jossey Bass, San Francisco, CA.

Wynne, B. (2002). Risk and Environment as Legitimatory Discourser of Technology: Reflectivity inside and out. *Current Sociology* 50(2):459–477.

Young, E. (1997). Dealing with hazards and disasters: risk perception and community participation in management. *Australian Journal of Emergency Management* 13(2):14–16.

Chapter 8

The Hazards Risk Management Process

Greg Shaw

Objectives

The study of this chapter will enable you to:

1. Establish the role/value of a hazards risk management process.
2. Define key terms associated with hazards risk management.
3. State the essential questions of hazards risk management.
4. Describe the Government Accountability Office framework for risk management and its inherent limitations.
5. Describe an overall framework for accomplishing the comprehensive hazards risk management process.
6. Explain the content and importance of each component within the hazards risk management framework.
7. Explain how the hazards risk management process supports comprehensive emergency management.

Key Terms

Comprehensive emergency management
Hazards

Hazards risk management
Management
Prevention
Risk
Risk analysis
Risk assessment
Risk communication
Stakeholders

Issue

What process will best inform decision makers in their efforts to balance safety and security expenditures with the myriad challenges, requirements and opportunities facing all organizations and communities?

Critical Thinking: Billions of dollars are spent in organizations from all sectors (private, public, and not-for-profit) and all levels of community, from individuals and their families to the federal government, on measures to manage risk from natural, technological, and intentional hazards. Perfect hazard risk management is unobtainable, and decisions must be made to consider and formulate hazard risk management interventions in the context of overall organizational/community priorities. As presented and explained in this chapter, can the hazards risk management process inform decision makers in establishing priorities that balance competing needs while devoting limited resources to the most effective and efficient risk management interventions?

Introduction

Chapter 1 describes the nature, purpose, and application of hazards analysis as a process and a tool that supports the phases of comprehensive emergency management (preparedness, mitigation, response, and recovery). This chapter takes a step back from hazards analysis as an activity undertaken to understand hazards and the risks they pose. It focuses on the larger hazards risk management (HRM) philosophy and framework as an iterative and ongoing process that is intended to inform decisions dealing with safety, security measures, and sustainability at all levels of organizations and communities. The structure of the HRM framework described in this chapter is adapted from the Emergency Management Australia emergency risk management process set forth in the *Emergency Risk Management Applications Guide*. A much more detailed description and discussion of HRM can be found in 1,000 plus pages of the Federal Emergency Management Agency EMI Emergency Management Higher Education Project Hazards Risk Management course available on the Higher Education Project Web Site (FEMA 2004).

Terminology

As discussed in Chapter 1, there are multiple, and often conflicting, definitions of terms associated with hazards and HRM. These definitions may change over time to reflect certain areas of emphasis and are not necessarily consistent, even within a particular discipline or organization.

For example, The National Fire Protection Association (NFPA) issued the 2004 document, *Standard on Disaster/Emergency Management and Business Continuity Programs*, which defines mitigation as "activities taken to eliminate or reduce the probability of the event, or reduce its severity or consequences, either prior to or following a disaster/emergency" (NFPA 2004: 4). The 2007 edition of this document redefines mitigation as "activities taken to reduce the severity or consequences of an emergency" (NFPA 2007: 4) and introduces the new term, prevention, which is defined as "activities to avoid an incident or to stop an emergency from occurring" (NFPA 2007: 5). Following from these definitions, mitigation, as a widely understood and accepted phase of the long-established framework of comprehensive emergency management, has been bifurcated into the two phases of prevention and the newly defined meaning of the term mitigation, which focuses on consequence management.

To complicate matters further, mitigation is defined in the Department of Homeland Security–issued National Response Plan of December 2004 in the more traditional manner as "activities designed to reduce or eliminate risks to persons or property or to lessen the actual or potential effects or consequences of an incident" (United States Department of Homeland Security 2004: 68). The thrust of this definition, which maintains prevention activities within mitigation, is retained in the Draft National Incident Management System of August 2007 (FEMA 2007: 21) and the Draft National Response Framework of September 2007 resource Web site.

Considering this example, differing definitions for the HRM process and terms contained within the process are to be expected and accepted. To avoid confusion, these terms should be defined and used consistently. Accordingly, the following terms related to the HRM process are presented and defined, along with the rationale for selecting the chosen definition for use in this chapter.

Lacking a widely accepted definition for the term HRM, the term is defined based upon its three component words: hazard(s), risk, and management.

Consistent with a definition of hazard included in Chapter 1, the definition from the 1997 FEMA publication *Multi Hazard Identification and Assessment* is selected for developing a definition of HRM: "Events or physical conditions that have the potential to cause fatalities, injuries, property damage, infrastructure damage, agricultural loss, damage to the environment, interruption of business, or other types of harm or loss" (FEMA 1997: xxv). Defining hazards in this manner is purposeful, since it is inclusive of all sources of hazards and does not necessarily emphasize any one category of natural, technological, or human-induced (intentional/terrorist) events.

Risk and the more expansive concept of risk management are also subject to multiple definitions and are often misunderstood or confused with other terms, such as risk identification, risk assessment, risk analysis, and risk communication. As discussed later in this chapter, risk management is a function comprised of several subfunctions that work together for the purpose of informing decision making at all levels of organizations and communities. Risk, as the foundational term for risk management, has differing meanings in different disciplines such as medicine, finance, safety, security, etc. The selected definition for risk derived from Ansell and Wharton (1992) is general in nature and applies across these disciplines. Risk is the product of probability (likelihood) and consequences of an event. Defining risk in this manner implies that risk can be managed by influencing or the probability (through mitigation and preparedness actions) and consequences of a disaster (through mitigation, preparedness, response, and recovery actions).

The chosen definition for manage comes from the *Merriam-Webster Dictionary*: "To work upon or try to alter for a purpose." Other definitions of manage include words like direct, govern, and succeed, which imply achieving control. Although a manager of risk strives to achieve control over risks, this is generally not totally achievable due to uncertainties, unknowns, and other intervening concerns. As stated by Borge, "Risk management is not, and will never be, a magic formula that will always give you the right answer. It is a way of thinking that will give you better answers to better questions and by doing so helps you shift the odds in your favor" (2001: 4). In dealing with risk, one is seldom or never in complete control, and the best one can do is work to influence future events in a manner that is perceived as favorable.

Therefore, combining these three definitions with the author's personal bias, HRM is defined as: A process that provides a general philosophy and a defined and iterative series of component parts that can be utilized to establish goals and objectives and inform decisions (strategic and tactical) concerning the risks associated with all hazards facing an organization and/or community. This definition of HRM is intended to emphasize each of the three component terms and the application of the process to all hazards and all phases of comprehensive emergency management. HRM, as an iterative process, is thus intended to provide an understanding of hazards and risks and a rational, inclusive, and transparent process for identifying, assessing, and analyzing hazard risks across all sectors and at all levels of community to inform decision makers as they allocate limited resources to the myriad and often competing priorities of their organization/community.

As discussed in the following section, risk management (a more commonly used term that can be used synonymously with HRM) has gained prominence in the post-9/11 environment, particularly as a tool for dealing with human-induced (intentional/terrorist) hazards. This predominantly terrorism-focused application of risk management has evolved to a more HRM all-hazards focus, particularly with the fallout from Hurricane Katrina and the perceived failures of all levels of government to adequately mitigate against, prepare for, respond to, and recover from the catastrophic events resulting from natural and technological hazards.

Risk Management

In the post-9/11 environment the term risk management has gained prominence, particularly in the vernacular and practice of Homeland Security. The Homeland Security Act of 2002 requires the Department of Homeland Security (DHS) to conduct comprehensive assessments of vulnerability (a component of risk) to the critical infrastructure and key resources of the United States (White House 2002). Homeland Security Presidential Directives (HSPD) 7: Critical Infrastructure Identification, Prioritization, and Protection, and 8: National Preparedness, both issued in December 2003, endorse risk management as a way of allocating resources (White House 2003A, White House 2003B). The National Infrastructure Protection Plan issued in July 2006 is based upon three foundational blocks including a "risk management framework establishing processes for combining consequence, vulnerability, and threat information to produce a comprehensive, systematic, and rational assessment of national or sector risk" (United States Department of Homeland Security 2006: 35). Within the *National Infrastructure Protection Plan*, Chapter 3 is titled "The Protection Program Strategy: Managing Risk," and Chapter 7, titled "Providing Resources for the CI/KR Protection Program," includes a section titled "The Risk-Based Resource Allocation Process."

The commitment to a risk management–based approach within DHS was further demonstrated by the newly appointed Secretary Michael Chertoff in the months following his confirmation. In his April 26, 2005, address to government and business leaders at New York University, Secretary Chertoff stated,

> Risk management is fundamental to managing the threat, while retaining our quality of life and living in freedom. Risk management can guide our decision-making as we examine how we can best organize to prevent, protect against, respond and recover from an attack . . . For that reason, the Department of Homeland Security is working with state, local and private sector partners on a National Preparedness Plan to target resources where the risk is greatest (Chertoff 2005).

Although terrorism focused, Secretary Chertoff's remarks can and should be extended to all hazards and clearly emphasize the importance of risk management in "guiding" decision making supporting comprehensive emergency management.

The experiences observed in the next year and a half and the lessons learned during the 2005 hurricane season only strengthened Secretary Chertoff's commitment to risk management as a foundation of Homeland Security. In his December 14,

2006, address at The George Washington University, Washington, DC, Secretary Chertoff stated,

> Probably the most important thing a Cabinet Secretary in a department like this can do as an individual is to clearly articulate a philosophy for leadership of the department that is intelligible and sensible, not only to the members of the department itself, but to the American public. And that means talking about things like risk management, which means not a guarantee against all risk, but an intelligent assessment and management of risk; talking about the need to make a cost benefit analysis in what we do, recognizing that lurching from either extreme forms of protection to total complacency, that's not an appropriate way to build a strategy; and finally, a clear articulation of the choices that we face as a people, and the consequence of those choices (Chertoff 2006).

Taken together, Secretary Chertoff's remarks, though separated by time and events by over 18 months, emphasize several very important points concerning the purpose and application of risk management:

1. Risk management can "guide" (inform) decision making across the phases of comprehensive emergency management.
2. Risk management is applicable to and across all levels of government (local, state, federal), all sectors (public, private, and not for profit) and to the American public.
3. Decisions based upon risk management should include a cost–benefit analysis (not just monetary costs and benefits but all costs and benefits such as social, political, public relations, etc.)
4. Communication (clear articulation) is a necessary component of risk management.
5. Risk management should support strategic planning and management.

Critical Thinking: Stephen Flynn, in his 2004 book *America the Vulnerable*, makes the very profound statement concerning understanding and dealing with risk in the post-9/11 environment: "What is required is that everyday citizens develop both the maturity and the willingness to invest in reasonable measures to mitigate that risk" (Flynn 2004: 64). How do we, as everyday citizens, gain this maturity and willingness to understand the risks facing our organizations/communities and to accept as reasonable the measures taken to mitigate that risk? What are our roles and responsibilities as members of our organizations and communities to engage in a process for managing risk, and what are the roles and responsibilities of our organizational and community leaders to include us in that process?

To address these key points, a widely distributed, understood, and accepted framework for risk management is needed. Recognizing this need, the Government Accountability Office developed and distributed a Risk Management Framework displayed in Figure 8.1 (Government Accountability Office 2007: 9).

The GAO report from which this framework was extracted makes the point that, "Risk management, a strategy for helping policymakers make decisions about assessing risks, allocating resources, and taking actions under conditions of uncertainty, has been endorsed by Congress and the President as a way to strengthen the nation against possible terrorist attacks" (Government Accountability Office 2005: 5). The report goes on to state, "GAO developed a framework for risk management based on industry best practices and other criteria" (Government Accountability Office 2005: 6). This framework, shown in Figure 8.1, divides risk management into five major phases: (1) setting strategic goals and objectives, and determining constraints; (2) assessing the risks; (3) evaluating alternatives for addressing these risks; (4) selecting the appropriate alternatives; and (5) implementing the alternatives and monitoring the progress made and results achieved.

Given that the GAO has provided an authoritative and relatively widely accepted framework and approach to risk management, why is an alternative HRM framework and process required? The GAO framework as presented is in fact inclusive of certain components of the HRM process, but goes beyond the intent of HRM to include risk-based decision making and the implementation and monitoring of these risk management decisions. The HRM process, as described in the following sections, provides a context for risk-based decision making and the identification,

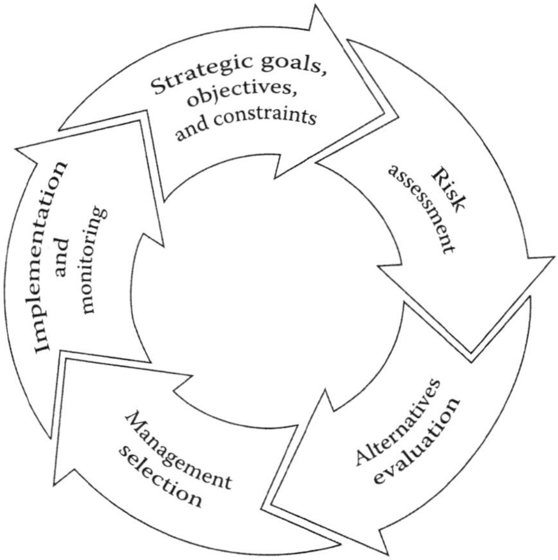

Figure 8.1 Risk management framework.

assessment, analysis, and presentation of hazard risk data and information. HRM is intended to support comprehensive emergency management as one input to informed decision making that attempts to balance safety and security expenditures with the myriad challenges, requirements, and opportunities facing all organizations and communities. The GAO framework also implies that the component steps are sequential, which they are not. The steps influence each other throughout the process and later steps may necessitate the revisiting of earlier steps and revisions of the results of each step.

A major shortfall of the GAO framework is that it largely ignores the necessity of *continuous* risk communication and monitoring and review throughout the overall process, which can doom the overall process to failure. The point of emphasis here is that HRM is an ongoing process that continually examines the impact of organizational activities to ensure that risks are identified, considered, and understood to support decisions impacting our vulnerability to those risks. To maximize effectiveness, any risk management process must continuously communicate strategies and tactics to manage the adverse impacts of risks throughout the impacted organization/community.

To improve the risk management process, a set of framing questions and a framework for HRM are presented and described as a recommended philosophy and approach to informing safety and security decision making in any sector and at all levels of organizations and communities.

Hazards Risk Management Framing Questions

Before embarking on the HRM process, and particularly before starting any risk assessment, the following questions should be asked and answered in a manner generally understood and acceptable to the audiences impacted by the HRM process results.

What are the organization's/community's strategic goals and objectives, and considering those goals and objectives:

- What is the scope of our hazards risk management effort?
- What is an acceptable level of risk?
- Who determines what an acceptable level of risk is?
- Can risk be managed?
- What are the interventions (controls/countermeasures) available to manage risk?
- What combination of risk management interventions (controls/countermeasures) make sense in terms of non-risk-specific considerations (economic, social, political, legal)?

Framework for Hazards Risk Management

Figure 8.2 displays the hazards risk management framework as adapted from the emergency risk management process set forth in the 2002 Emergency Management Australia, *Emergency Risk Management Applications Guide* (Emergency Management Australia 2004). The HRM framework includes the general format of the emergency risk management framework but meets a different purpose, as described in this section. The HRM framework includes six steps: (1) establish the context, (2) identify the hazards, (3) assess the hazards risk, (4) sort the hazards by risk magnitude, (5) analyze the risks from each hazard, and (6) group and prioritize risks; and two continual components: communicate and consult, and monitor and review. Roughly categorized, steps 1 and 2 accomplish hazard identification, steps 3 and 4 hazard risk assessment, and steps 5 and 6 hazard risk analysis. Note that chapters in this book examine hazard identification and characterization, modeling, spatial analysis, risk, and vulnerability analysis. We thus view the hazards analysis process

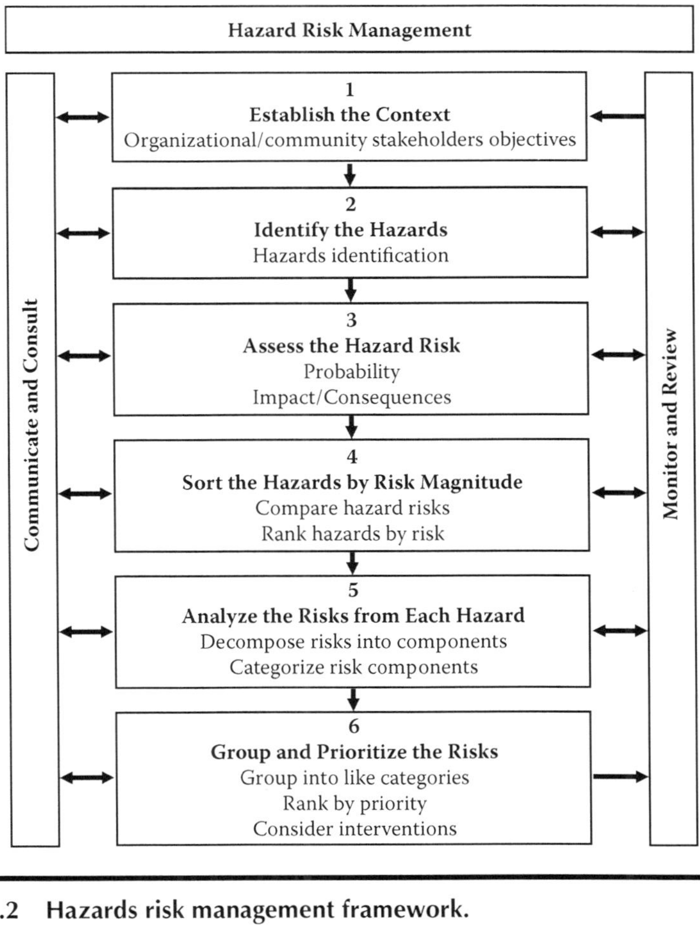

Figure 8.2 Hazards risk management framework.

in the context of hazards risk management and as a process to generate information for selecting appropriate hazard mitigation strategies.

The HRM framework is constructed to define an inclusive, iterative, and continuous process that addresses the HRM framing questions listed above and provides a foundation for the four phases of comprehensive emergency management (CEM): preparedness, mitigation, response, and recovery. Inherent in each of the phases of CEM is the goal of effectively and efficiently managing the myriad hazards that may adversely impact an organization/community and its ability to achieve its strategic and tactical goals and objectives. Following the HRM process is intended to provide the "needs assessment" for comprehensive emergency management, and as such, establishes a focus and steering direction. Understanding the HRM process is a key to developing a risk-based, all-hazard emergency management program.

Each component of HRM Process is discussed in this section of the chapter. Much of the content in this discussion is adapted from *Emergency Management Principles and Practices for Healthcare Systems*, authored by The Institute for Crisis, Disaster, and Risk Management (ICDRM) at the George Washington University (GWU) for the Veterans Health Administration (VHA)/US Department of Veterans Affairs (VA). Washington, DC, June 2006 (Barbera et al. 2006).

Components of the Hazards Risk Management Process

Communicate and Consult (A Continual Component of the HRM Process)

Continual communication and consultation within and without an organization/community provides a means of inclusion and the establishment and management of realistic expectations for the HRM process and its eventual incorporation into the organization's/community's overall emergency management program. Step 1 in the HRM process calls for establishing the organizational/community context, involving stakeholders, and setting objectives. Communication and consultation does not stop there. Repeating the statement on risk management of Secretary Chertoff of DHS from earlier in this chapter, he views his responsibilities as a Cabinet Secretary to include the need to "clearly articulate a philosophy for leadership of the department that is intelligible and sensible, not only to the members of the department itself, but to the American public" (Chertoff 2006). This role is shared by all leaders of organizations and communities with risk management responsibilities. To meet this responsibility the entire HRM process should be open, accessible, and intelligible to the impacted public.

As a matter of guidance and emphasis for continuous communication and consultation throughout the HRM process, the 1989 National Research Council Report, *Improving Risk Communication*, provides the following statement that should guide all risk communication. "Risk communication is a process, the success of which is measured by the extent that it, first, improves or increases the base of information that decision makers use, be they government officials, industry managers, or individual citizens, and second, satisfies those involved that they are adequately informed within the limits of available knowledge" (National Research Council 1989: 74).

It should also be noted that we stress the role of communication and stakeholder participation in the hazards analysis process in both Chapter 1 and 7 of this book. The public has a critical need to know and understand the nature of risks in the community, and risk communication should be an intentional part of hazard risk management and the hazards analysis process.

Monitor and Review (A Continual Component of the HRM Process)

The HRM process is never actually finished, as it is subject to reanalysis and revision when changes occur in the internal and external environments. Continuous monitoring and review of findings from all steps should be conducted to keep the overall process relevant and on track with the emergency management program. Drills, exercises, and actual events will test the emergency management program, and both the positive and negative observations related to system vulnerabilities should be noted and analyzed. The HRM process also constitutes a major means of monitoring and reviewing any findings related to reduced as well as to newly recognized hazard risks. For example, an exercise could examine whether a new process or procedure has effectively reduced a previously recognized hazard risk. Similarly, an exercise, a threat, or an actual event may prompt the recognition of previously unidentified hazards and/or hazard risk.

Step 1: Establish the Context

Context refers to the external environment in which the organization/community exists and functions and the internal characteristics of the organization/community itself. Therefore, establishing the context for the HRM process (and, essentially, for an overall emergency management program), is the logical starting point for the process. To accomplish this, the organizational/community context, the stakeholders, and the objectives for the HRM process must be defined.

Organizational/Community Context

The organizational/community context for the HRM process is established based upon the organization's/community's responsibilities, the social, economic, political, and legal realities, and the review and input of stakeholders. This step in the process begins hazard identification, where an organization or community is characterized from a social, economic, political, geographical, structural, and ecological context. In short, the starting point of HRM is to clearly describe the nature of our organization/community.

The organization's/community's boundaries, to include strategic and tactical goals and objectives and legal and moral roles and responsibilities, are delineated. Additionally, economic, social, political, and legal constraints on the organization/community for resource allocation supporting emergency management requirements and initiatives should be identified and recognized. This step helps to answer all framing questions. For example, a community should consider its demographics, economic state, strategic and tactical goals for growth and development, and its roles and responsibilities to its population and to surrounding communities to determine the scope and constraints of its HRM effort.

Stakeholder Involvement

Identifying and engaging a stakeholder group is the second critical step in establishing the context and particularly can assist in identifying the social, economic, political, and legal realities and constraints that impact the HRM process. The individual/group responsible for the HRM process (HRM Committee) accomplishes this by identifying all appropriate stakeholders, both internal and external to the organization/community, that should be included and considered in all steps of the HRM process. Stakeholders are defined as key people, groups of people, or institutions that may significantly influence the success of the process.

Stakeholder analysis is a technique that is increasingly employed in private industry to identify and assess the importance of stakeholders and thereby judge that the stakeholder group is balanced and comprehensive. To ensure that multiple perspectives are adequately considered and represented in the overall HRM process, the following steps help define a successful stakeholder analysis (Management Sciences for Health and United Nations Children's Fund 1998):

- Identify people, groups, and institutions that will influence your HRM process.
- Develop strategies to build the most effective support possible for the process and reduce any obstacles to successful implementation of an effective emergency management program. For example, simply inviting outsiders such as representatives from business, public safety, local emergency planning committee, and impacted community groups into the HRM process may help resolve misconceptions and miscommunications.

Objectives

Establishing the specific objectives for the HRM process follows from defining the organizational/community context, and the involvement of the appropriate stakeholders. Realistic and measurable objectives based upon observable outcomes for both strategic (long-term) and tactical (short-term) activities are essential for all other steps and components of the HRM process, particularly communication and consultation, and continuously monitoring and reviewing the process and the results.

Step 2: Identify the Hazards

This component involves the listing of all possible hazard types that could significantly impact the organization/community.

Comprehensive Hazard Identification

The full range of hazards must be captured. The list includes hazards that do not directly or physically impact the organization/community, but could generate demand on the organization's/community's resources from other organizations/communities. Hazard identification should also include hazards that, if they occur, could cause major impacts on the goods and/or services provided by the organization/community such as loss of customers, litigation, liability payments, poor publicity, etc. A comprehensive examination of the hazard identification process is provided in Chapter 2.

Hazard Identification Strategy—Organization/Community Resources

While multiple resources are available to assist in identifying hazards, an essential consideration is coordination with outside organizations/communities, including nearby organizations and municipalities, regional leadership, and state authorities. Hazards that could potentially impact an organization/community are commonly hazards that may impact the larger area and, therefore, have already been identified or are being defined by other organizations/communities at the local, regional, and state level.

Hazard Identification Strategy—Web Resources

Other resources are available to assist in hazard identification. These include local, state and national Web sites, FEMA and NOAA publications, and the 15 National Preparedness Scenarios established within the 2007 *National Preparedness Guidelines* (DHS 2007). Examples include: the Washington, DC, Emergency Management Agency provides a list and description of the "18 Major Hazards" expected to impact the DC area (DC Homeland Security and Emergency Management Agency 2007); the State of Virginia Department of Emergency Management provides a list and description of potential hazards (Virginia Department of Emergency Management Agency 2008); FEMA provides historical data and links to hazard-related Web sites (FEMA Web site). Many of these resources provide important historical data on hazards.

Categorizing Hazards

As hazards are identified, it is useful to group the hazards according to the following categories, where commonalities predominate both in cause and in actions necessary to address the hazard risk. These categories are further explored as part of the hazard identification process within a hazard modeling context in Chapter 3.

- *Natural hazards:* Hazards that primarily consist of the forces of nature
  - For example, hurricane, tornado, storm, flood, high water, wind-driven water, tidal wave, earthquake, drought, lightning-caused wildfire, infectious disease epidemic
- *Technological hazards:* Hazards that are primarily caused by unintentional malfunction of technology, including human and system actions
  - For example, industrial, nuclear, or transportation accidents; power and other utility failure; information technology failure; hazardous materials release; and building collapse
- *Intentional hazards:* Hazards that are caused primarily by deliberate human threat or executed action. These are usually criminal, civil disobedience, or terrorist in nature
  - For example, civil strife, terrorism, or criminal attacks on the community

Step 3: Assess the Hazard Risk

Hazard risk assessment is conducted in the next two HRM steps. Risk, as previously defined in this chapter, is a product of probability and consequences. Each hazard identified by the organization/community should, therefore, be assessed individually according to its probability of occurrence and its impact (consequences) on the organization/community as a means of approximating each hazard's level of risk.

Hazard Risk Assessment Strategy

How the hazard risk assessment is presented and accomplished varies among sources, but all share the common purpose of establishing the relative importance of and between hazard risks. For that reason, most hazard risk methods employ a ranking system that assigns a quantitative value to each individual hazard to allow a preliminary method of sorting by numeric value. It must be remembered, however, that the assignment of quantitative values to probability and consequence is often subjective and can be based upon information with inherent uncertainties.

For example, the exact probability of any event occurring as a result of a natural, technological, or intentional hazard is not necessarily determined by past occurrences and is subject to continual changes in the organization's/community's internal and external environments. Determining the probability of an intentional hazard (terrorist strike) impacting an organization/community is particularly perplexing due to the general lack of historical data and the fact that an intelligent hazard vector, such as a terrorist, can adjust the location and nature of the attack based upon the hazard risk management controls in place.

Similarly, an exact determination of consequence is not possible, since it requires an a priori understanding of the hazard event scenario prior to the occurrence and a complete understanding of the organization/community impacted. Complicating the matter further is the necessity to define the categories (people, property, vital services, business loss, etc.) to be included in the measure of consequence and a common metric for combining the categories. Also, looking beyond the immediate consequences of an event, long-term consequences such as environmental damage and financial impacts of a localized event may permanently alter the environment and cascade through the larger economy. These long-term consequences may dwarf the immediate consequences and are not necessarily quantifiable (e.g., the immediate consequences on a maritime container port in terms of physical damage and loss of life due to a hurricane, as compared to the longer-term impact on the environment, local economy, and possibly the overall national economy).

For these reasons it is essential that the purpose of the hazard risk assessment be emphasized throughout the HRM process—to establish the relative importance of hazard risk between the identified hazards for sorting—and that the stakeholders providing input to the assessment have a common understanding of the rating measures in order for them to be consistent across all inputs. A discussion of the myriad hazard risk assessment methods and templates currently in use go far beyond the scope of this chapter. The key point is that the method and template selected is useful and appropriate for the scope of the particular HRM effort.

Step 4: Sort the Hazards by Risk Magnitude

This step continues the hazards risk assessment and consists primarily of assigning a relative level of importance to each hazard risk value from Step 3, thereby placing each hazard risk in the context of the overall cohort of the identified hazard risks.

Sorting Strategies

This sorting of hazard risks entails the comparison of the assigned risk values (established in Step 3) associated with each hazard and the designation of each hazard to one of the broad categories (high risk, moderate risk, and low risk) via mathematical or expert judgment methods. While simple numerical values are commonly used to represent probability and consequence, the comparative value of the selected metrics must be fully understood for this ranking system to have merit. In other words, is a value of 3.92 calculated in the risk assessment essentially equal to a value of

4.15? If so, separating hazards with these two values into different categories may be misleading and cause judgment errors later in the HRM process. To address this, presenting the score for all identified hazards on a single graph or spreadsheet may allow appropriate grouping of hazards, with less reliance on specific (but relatively arbitrary) numerical assignments.

The Use of Expert Judgment

Expert judgment should enter into this hazard sorting and may result in a rearrangement of the results based upon specific intelligence related to probability and/or consequence. For example, a terrorist attack using biological agents may have almost unimaginable consequences that totally dwarf the probability considerations, thus elevating such an event to the top of the priority list. In the absence of specific intelligence, however, the rank of this hazard may be moved to below that of an event with a better-defined probability, such as a hurricane in a rural coastal community.

Conversely, a lower-ranked terrorist hazard may suddenly be elevated to the top of the hazard list based upon new threat information, thereby overriding the earlier expert judgment. The dynamic nature of the natural, technological, and intentional hazard environment necessitates an expert-level review and judgment beyond mere numerical sorting. A complicating factor in the use of expert judgment is the definition and identification of true "experts." In the post-9/11 environment, there appear to be many "terrorism experts," ready and willing to express their views on the topic. What exactly qualifies them as experts may be subject to debate.

Step 5: Analyze the Risks from Each Hazard

Hazard risk analysis is accomplished in the next two steps. Hazards are considered individually during the earlier steps of the HRM process, and the risk (probability and consequences) of each hazard is compared only at a very macro level. The final analysis step of the HRM process should allow decision makers to look across all hazards to identify components of hazard risk that are common to multiple hazards. At this point, the components of risk refer to the consequences of an event resulting from the occurrence of a hazard. This approach promotes the identification of options that reduce or eliminate components of risk from multiple hazards through a single intervention (see Step 6) and therefore supports the most effective and efficient allocation of resources to HRM.

Hazards Risk Analysis Strategy

To accomplish this, the components (consequences) of risk from each hazard should be "decomposed" into significant elements that can be compared and/or grouped across the range of identified hazards. A relatively common grouping of hazard risk consequences is in the categories of human, property, and operational consequences.

The groupings should refer to the processes and resources that are disrupted (i.e., so that they can later be grouped across hazards according to the "all-hazard" processes and resources that are affected). For example, a generic organization experiencing a hurricane would expect to experience consequences in each category with specific consequences that impact the resources and operational processes of the organization. The specific consequences and disruptions could include:

- Human consequences:
  - Inability for staff to reach, or remain at, work: child/elder care responsibilities, transportation disruption, concern about personal property, loss of personal property, and others
  - Injury/death to staff (at work or at home) and customers/clients/guests within the facility due to high winds and debris causing window/door glass failure
- Property consequences:
  - Flooding, roof failures, and other water effects
  - Wind and debris damage to buildings, outside equipment, vehicles, and other property on facility premises
  - Storm surge effects if relevant
  - Maintenance problems due to failure of personnel to report for work
- Operational consequences:
  - Inability to provide products and/or services
  - Inability to meet contractual agreements
  - Interruption of cash flow
  - Damage to reputation

Categorizing the Hazard Risk

For each significant hazard, the hazard risk is analyzed, decomposed into elements, and grouped in a format that will allow like elements to be identified across all hazards.

Step 6: Group and Prioritize the Hazard Risks and Consider Risk Management Interventions

This step completes the hazards risk analysis by sorting and comparing the hazard risk elements determined in Step 5.

Grouping and Prioritizing Strategy

It is very likely that some identical hazard risk elements (defined in Step 5) will be present in a wide range of hazards that cross the natural, technological, and intentional hazard categories. For example, nearby hazardous materials releases with explosive potential (technological or intentional), an approaching tornado (natural), and a realistic truck bomb threat (intentional) all expose an organization's employees, staff, and guests to injury and death due to the physical damage to the building unless prevented/mitigated by structural measures, immediate protective actions, and the availability of an emergency response capability. Guarding against this hazard risk element (injury and death) would be prioritized with other hazard risk elements according to a prioritization scheme which generally places life and safety issues first, with property protection and continuity of operations as lower priorities.

Consider Hazard Risk Element Interventions

The individual hazard risk elements are further analyzed to develop potential interventions for consideration in the development of formal hazard mitigation and preparedness plans. Following from the above example, some potential interventions could include:

- Mitigation/prevention interventions to reduce the risk of injury or death by *reducing the likelihood* of physical damage to the building:
  - Structural measures to increase the strength of the building
  - Removing windows
  - Covering windows with protective coatings
  - Relocating personnel work spaces away from outer walls
  - Standoff barriers to keep vehicular traffic away from the building's perimeter
  - Building security measures to check vehicles entering the parking area
- Mitigation/consequence management preparedness interventions to reduce the risk of injury or death *by reducing the consequences* of physical damage to the building:
  - Emergency action plans covering sheltering in place and evacuation
  - Communication capabilities (e.g., general announcement, alarms, computer alert) that deliver relevant and actionable information to building occupants
  - Communication capabilities that receive and deliver relevant and actionable information from and to outside public safety organizations
  - Awareness, training, and exercises
  - Internal emergency response organization and capability
  - Mutual aid agreements with other organizations/communities
  - Coordination with public safety organizations

If the identified interventions are applicable across multiple hazards, this may prompt further grouping of risk elements. At this step in the process, the risk elements are grouped together. *Selection of interventions for implementation, however, is an activity that occurs later, during formal emergency management planning.*

Consideration of potential interventions is used only to prompt the grouping of risk elements in a manner where they may be addressed through economy of scale or in a manner that provides greater benefit than if each element is individually addressed.

Obviously, each potential intervention has resource implications (costs—both tangible and intangible) that must be considered in the context of contribution (benefits) to the HRM goals and objectives. In general, costs, particular monetary and time costs, are easy to quantify. However, intangible costs such as reduced employee morale, or decreased accessibility to a building for employees, customers, and guests, disruption of previous policies and procedures may be significant and should not be ignored. Actual benefits derived from risk interventions are much more difficult to identify and quantify. Particularly in the absence of a hazard event, benefits may be largely invisible to decision makers who must allocate limited resources to multiple organizational/community priorities that may have little or nothing to do with HRM.

Clearly, some of the preparedness interventions such as planning, awareness, training, exercising, coordination and mutual aid are of relatively low cost, can be implemented easily, and provide some contributions to achieving HRM goals and objectives. Others, such as internal and external communication and the development of internal response organization and capability, have higher costs, but are recognized as applicable across all hazards, and as such, may be judged as cost effective.

Application of the Hazards Risk Management Process

Overview

An example of processing an identified hazard through the six-step HRM process may serve to illustrate the value of this approach. The organization/community selected for this example is a fictional private urban university that is both an organization (a not-for-profit business) and a distinct community within the larger urban city community (8,000 resident students, 12,000 nonresident students, 8,000 faculty and staff, and a capital plant valued in excess of one billion dollars). The university can be impacted across the spectrum of hazard categories (natural, technological, and intentional) as described in Step 2 of the HRM process. *One identified hazard, civil strife, is selected for this example.* The university is located in proximity to highly visible and controversial organizations/entities that are subject to protest events coinciding with scheduled and widely publicized meetings. Previous protest events have been marked by violence, injuries, property damage, disruption of local transportation, restricted access to areas of the community, and arrests of protesters and bystanders.

Step 1: Establish the Context

Organizational/Community Context

The university's context is described in terms of its stated strategic goals and objectives.

A review of the university's strategic plan provides the overall goal of "providing a safe and supportive environment for students, staff, and faculty to pursue education, research, and personal and professional growth." Supporting this goal are specific objectives:

1. Maintaining the safety and well-being of all members of the university's community.
2. Protecting the physical and intellectual property of the university.
3. Maintaining continuity of essential university operations following disruption.
4. Preserving the university's reputation as a high-quality institution of higher learning.
5. Supporting the surrounding community and the overall urban community as a good neighbor.

For any and all hazards, the university is focused on the above listed objectives in the order of priority as listed. The highest priority is to afford protection for all members of the university community. The HRM process should look across all of the phases of comprehensive emergency management (mitigation, preparedness, response, and recovery) to assess and analyze the hazard risks and to identify possible hazard risk management interventions for the university's leadership consideration. The safety and well-being of community members demands a primary focus on mitigation (prevention and consequence management) interventions.

The university's physical location, legal and moral roles, responsibilities, and authority determine the scope and constraints of the HRM effort. For the example hazard, civil strife, reduction of risk to a zero level is not obtainable short of removing the hazard, prohibiting the members of the university community from coming into any contact with the hazard, and/or cordoning the university off from the hazard. None of these measures are possible due to the physical setting of the university, the commitment of the university to academic and intellectual freedom, and the rights of the university and surrounding community members. Therefore, there is some level of risk associated with the example hazard, and the university can employ the HRM process to establish an acceptable level of risk and identify the potential interventions to be considered for implementation.

Stakeholder Involvement

Stakeholders include representatives of the university's student body (all categories of students including resident and nonresident and undergraduate and graduate), faculty, staff, students' parent groups, alumni, police, fire department, urban government authorities, surrounding communities, other area universities, etc. The level of involvement of each stakeholder group depends on the nature of the identified hazards as the HRM process proceeds. For the example hazard, all of these stakeholders have some level of involvement with public safety and urban government authorities as the primary source of relevant information and required coordination.

Objectives

Establishing the specific objectives for the HRM process follows from defining the organizational/community context and the involvement of the appropriate stakeholders.

Step 2: Identify the Hazards

This component involves the listing of all possible hazard types that could significantly impact the organization/community. For this example, civil strife is identified as a primary hazard. A comprehensive hazard identification strategy extending beyond this example would include research, expert consultation, and stakeholder input and review to identify all hazards that may impact the university as an organization and as a community.

Step 3: Assess the Hazard Risk

Identified hazards are considered individually to determine their probability and impact for the purpose of assigning a level of risk.

Hazard Risk Assessment Strategy

For the example hazard, the assigned probability is very high and approaching certainty based upon the projected schedule for meetings in the future and the history of demonstrations for past meetings. On a numerical scale of 1 to 5 of probability ranging from highly unlikely (rating of 1) to almost certain (rating of 5), the example hazard would be assigned a probability rating of 5.

Revisiting the university's strategic objectives identified in Step 1, the consequences of civil strife can have significant immediate and long-term consequences impacting each of the objectives. Members of the university community may be injured, university physical property may be damaged, and university operations may be disrupted. Based upon these consequences and the university's perceived actions across the phases of comprehensive emergency management, the university's reputation and stature in the larger community may be damaged, impacting future enrollments, faculty recruitment and retention, and research funding. For these reasons, the consequences of civil strife are assigned a rating of 5 on a numerical scale from minimal (rating of 1) to significant (rating of 5).

These ratings are combined by a selected mathematical algorithm (usually multiplying or adding) to establish the relative level of hazard risk between the identified hazards for further sorting. In this example, any method of combining the probability and consequences would lead to a determination that civil strife poses a high level of hazard risk to the university.

Step 4: Sort the Hazards by Risk Magnitude

During this step a relative level of importance is assigned to each hazard risk value determined in Step 3 for the purpose of placing each hazard risk in the context of the overall cohort of the identified hazard risks. The numeric ratings of probability and consequence and their combination are subject to uncertainties and potential rating bias. For that reason the hazards risks should be sorted into general categories such as low, moderate, and high risk. The civil strife hazard would obviously be placed in the high risk category, while other hazards such as a hurricane might be placed in the moderate risk category, and an accidental spill of a toxic chemical in the low risk category.

In addition to sorting by mathematical ratings, expert judgment should be applied to the sorting process to account for the dynamic nature of hazard risk and specific intelligence related to the probability and/or consequences of the hazard risk. In this example, the meetings resulting in demonstrations are scheduled for specific dates, and outside of those dates the probability of civil strife declines to an unlikely or highly unlikely level. Additionally, intelligence may indicate changes in the size and motivation of the demonstrations for different meetings, which could change the consequences rating. These considerations can and should rearrange the sorting of hazard risk based upon the current situation and the available information.

Step 5: Analyze the Risks from Each Hazard

In this and the following step, each hazard risk is considered in the context of all identified hazard risks to determine commonalities across multiple hazards. For the example of civil strife, the components (consequences) of the hazard risk are decomposed into significant elements that can be compared and/or grouped across the range of identified hazards. Considering human, property, and operational consequences civil strife could result in:

- Human consequences:
  - Injury/death to students, faculty, staff, visitors, and transients at the site of the demonstrations and within the confines of the university due to localized or spreading violence
  - Detention/incarceration of students, faculty, and staff in the proximity of demonstrations

  - Inability of students, faculty, and staff to reach or remain at the university due to transportation disruptions or physical security measures
- Property consequences:
  - Damage to university buildings from the outside due to vandalism
  - Damage to university buildings from the inside due to access to the buildings and vandalism
  - Damage to other university property such as buses, signage, and outside areas due to vandalism
  - Maintenance problems due to inability of personnel to report for work
- Operational consequences:
  - Inability to conduct classes
  - Inability to conduct scheduled events
  - Inability to receive and/or ship supplies
  - Damage to reputation

Step 6: Group and Prioritize the Hazard Risks and Consider Risk Management Interventions

For each significant hazard, the decomposed hazard risk elements are grouped in a format that will allow like elements to be identified across all hazards. For example, a hurricane would share many similar hazard risk elements with civil strife such as the potential for injury/death, inability to reach or remain at the university, damage to university buildings and property, inability to conduct classes, conduct scheduled events, and receive and/or ship supplies.

Grouping and Prioritizing Strategy

The grouped hazard risk elements are then prioritized with life and safety issues first and with property protection and continuity of operations as lower priorities.

Consider Hazard Risk Element Interventions

The individual hazard risk elements are further analyzed to develop potential interventions for consideration in the development of formal hazard mitigation, preparedness, response, and recovery plans. For the civil strife hazard example some (for the sake of this example the list of interventions is not intended to be exhaustive) potential interventions include:

- Mitigation
  - Establishing policy that prohibits university community members from entering specified areas where civil strife is expected.
  - Cordoning off the university area with police enforcement.

- Limiting access to university buildings (dorms, classroom and office buildings, and administrative buildings) to only university community members with valid identification cards.
- Closing common buildings that are normally open to the public (student union, food courts, and athletic venues).
- Implementation of instant messaging technologies for connectivity with all university community members.
- Training/education for university community members on the nature of the hazard.
- Information and updates passed to university community members and stakeholder groups (particularly the parents of resident students) encouraging hazard avoidance and what to do if involved in the consequences of the hazard.
- Shutting down university operations during the expected period of the civil strife.

- Preparedness
  - Review of mitigation, preparedness, response, and recovery plans.
  - Training for personnel with mitigation, preparedness, response, and recovery responsibilities.
  - Liaison with public safety agencies and community emergency management agencies.
  - Information updates to all university community members and stakeholders as appropriate.
  - Increased medical response capabilities on campus.
  - Increased police patrols on campus.
  - Increase levels of essential supplies (food, water, medical).
- Response
  - Activate the university emergency response organization and plans.
  - Continue information updates to all university community members and stakeholders as appropriate.
  - Shut down university operations consistent with the situation.
- Recovery
  - Resume operations as the situation permits.
  - Assess and deal with the consequences of the event.
  - Provide counseling services for university community members as needed.
  - Communicate with all university community members and stakeholders as appropriate.
  - Conduct an after event review to capture lessons learned and to identify corrective actions.
  - Review and revise all emergency management–related plans, policies, and procedures based upon the after-event review.

Obviously many of the potential interventions such as planning, community-wide training and education, passing information, instant messaging technologies, and liaison with public safety and emergency management agencies are applicable across all categories of hazards and should be considered in the context of this widespread utility. Other interventions such as shutting down operations, increased police presence, increased medical response capabilities, and increased university building security, may be particular to the specific hazard or a limited group of hazards. Some interventions such as prohibiting university community members from being in the area of the civil strife and cordoning off the university may be highly effective for hazard risk deduction processes, but are not feasible within the scope and constraints of the HRM effort as described in Step 1.

This concludes the specific steps of the HRM process. The outputs of the process, including the hazard risk groupings and priorities and potential interventions, are communicated to the university's decision makers for their consideration in determining what can and will be done to manage hazard risks. The HRM outputs are an essential input to informed decision making that attempts to balance safety and security expenditures with the myriad challenges, requirements, and opportunities facing the university.

Hazards Risk Management and Comprehensive Risk Management

The true measure of utility and success of the HRM process is its application to informing decisions within the organization's/community's overall emergency management program. The HRM process serves to identify hazard risk management interventions across the phases of comprehensive emergency management: mitigation, preparedness, response, and recovery. From that point on, it is up to the decision makers to answer the last of the HRM framing questions to decide what combination of risk management interventions (controls/countermeasures) make sense (economic, social, political, legal) for their organization/community. As stressed throughout this chapter, HRM as a process and its component steps are not the sole input to answering this question. They are just one input to informed decision making that attempts to balance safety and security expenditures with the myriad challenges, requirements, and opportunities facing all organizations and communities.

Discussion Questions

What is the real value of the hazards risk management process to you as an individual and as a member of an organization and/or community?

What are the measures of success of the hazards risk management process?

As quoted from the 1989 National Research Council report *Improving Risk Communication:* "Risk communication is a process, the success of which is measured by 'the extent that it, first, improves or increases the base of information that decision makers use, be they government officials, industry managers, or individual citizens, and second, satisfies those involved that they are adequately informed within the limits of available knowledge." Do you agree with this statement, particularly as it applies to risk communication in the post-9/11 environment? Do you feel that the risk information available to you in the context of your personal and professional related responsibilities is adequate?

How could you increase your involvement in the HRM process for your organization/community?

Applications

Consider an organization you are affiliated with or your community (local, state, or federal), and apply the HRM process to identify hazard risk management interventions that should be considered. To assist in following the HRM process, answer the following questions.

1. Are there non-risk-specific considerations (economic, social, political, legal) that should be considered when setting goals and objectives for hazards risk management?
2. What are the goals and objectives for the hazards risk management process?
3. Who are the stakeholders that should be included in the hazards risk management process? Are there varying levels of inclusion for different stakeholders?
4. What is an acceptable level of risk, and who determines what is an acceptable level of risk?
5. What sources of information would you consult to identify the hazards that may directly or indirectly impact your organization/community?
6. How would you assess (quantify) the risk of the hazards identified? How would you sort the hazards by the risk magnitude?—Based upon your personal knowledge of your organization/community and available information (e.g., information from your local/state emergency management agency Web sites), assess representative hazards by their probability and consequences.
7. How would you analyze the hazard risk from the identified, assessed, and sorted hazards?
8. How would you group and prioritize the hazard risk from the identified, assessed, sorted, and analyzed hazards?
9. How would you present (communicate) the results of the HRM process to stakeholders and decision makers?

Web Sites

Department of Homeland Security. http://www.dhs.gov/index.shtm

Department of Veterans Affairs. Emergency Management (EM) Principles and Practices for Healthcare Systems—Emergency Management Strategic Health Care Group. http://www1.va.gov/emshg/page.cfm?pg=122

Emergency Management Australia. http://www.ema.gov.au/

Federal Emergency Management Agency. http://www.fema.gov

FEMA. FEMA: State Offices and Agencies of Emergency Management: Federal Emergency Management Agency. http://www.fema.gov/about/contact/statedr.shtm

FEMA. Higher Education Conference. Federal Emergency Management Agency, Emergency Mamagement Institute. http://training.fema.gov/EMIWeb/edu/

Government Accountability Office. http://www.gao.gov

National Fire Protection Association. http://www.nfpa.org

Ready America. Ready.gov—Prepare. Plan. Stay Informed. http://www.ready.gov/america/index.html

References

Ansell, J. and F. Wharton (1992). *Risk: Analysis, Assessment, and Management.* John Wiley and Sons, Chichester, New York.

Barbera, J., Macintyre, A., Shaw, G., Westerman, L., Seefried, V., and de Cosmo, S. (2006). *Emergency Management (EM) Principles and Practices for Healthcare Systems.* The Institute for Crisis, Disaster, and Risk Management (ICDRM) at the George Washington University (GWU). Washington, DC.

Chertoff, M. (2005). Remarks by Secretary of Homeland Security Michael Chertoff at the Center for Catastrophic Preparedness and Response and the International Center for Enterprise Preparedness, New York University. Release date April 26, 2005. http://www.dhs.gov/xnews/speeches/speech_0249.shtm

Chertoff, M. (2006). Remarks by Homeland Security Secretary Michael Chertoff on Protecting the Homeland: Meeting Challenges and Looking Forward. Washington, DC. George Washington University. http://www.dhs.gov/xnews/speeches/sp_1166137816540.shtm

District of Columbia Homeland Security and Emergency Management (2007). 18 Major Hazards, Retrieved December 18, 2007 from the DC HSEM Web site.

Emergency Management Australia (2004). *Emergency Risk Management: Applications Guide.* Manual 5. http://www.aesvn.org/resources/ema/manual_5.pdf

FEMA (1997). *Multi-Hazard Identification and Risk Assessment: The Cornerstone of the National Mitigation Strategy.* Federal Emergency Management Agency. http://www.fema.gov/library/viewRecord.do?id=2214

FEMA (2004). *Hazards Risk Management course, EMI Emergency Management Higher Education Project.* Federal Emergency Management Agency. http://training.fema.gov/EMIWeb/edu/

FEMA (2007). *Draft Revised National Incident Management System.* Federal Emergency Management Agency. http://www.fema.gov/emergency/nims/nims_doc.shtm

Flynn, S. (2004). *America the Vulnerable.* Harper Collins, New York.

Government Accountability Office (2005). *Risk Management: Further Refinements Needed to Assess Risks and Prioritize Protective Measures at Ports and Other Critical Infrastructure.* (Report No. GAO-06–91). http://www.gao.gov/

Management Sciences for Health and United Nations Children's Fund (1998). Quality guide—stakeholder analysis. http://erc.msh.org/quality/ittools/itstkan.cfm

NFPA (2004). *Standard on Disaster/Emergency Management and Business Continuity Programs.* National Fire Protection Association, Quincy, MA.

NFPA (2007). *Standard on Disaster/Emergency Management and Business Continuity Programs.* National Fire Protection Association, Quincy, MA.

National Research Council (1989). *Improving Risk Communication.* National Academy Press, Washington, DC.

United States Department of Homeland Security (2004). *National Response Plan.* (Publication NO. 0520-B). http://purl.access.gpo.gov/GPO/IPS56895

United States Department of Homeland Security (2007). *National Preparedness Guidelines.* http://www.dhs.gov/xprepresp/publications/gc_1189788256647.shtm

White House (2002). *The Homeland Security Act of 2002.* http://www.whitehouse.gov/deptofhomeland/bill/

White House. (2003a). *Homeland Security Presidential Directive 7: Subject: Critical Infrastructure Identification, Prioritization, and Protection.* http://www.whitehouse.gov/news/releases/2003/12/20031217-5.html

White House (2003b). *Homeland Security Presidential Directive 8: Subject: National Preparedness.* http://www.whitehouse.gov/news/releases/2003/12/20031217-6.html

Chapter 9

Planning for Sustainable and Disaster Resilient Communities

Gavin Smith

Objectives

The study of this chapter will enable you to:

1. Understand the nexus between the topics discussed in Chapters 1 through 9 and their application to hazard mitigation planning.
2. Understand the connection between the following concepts: hazards management, sustainability, disaster resilience, and planning.
3. Understand the hazard mitigation planning process.
4. Understand the role of planners, the plan-making tools and participatory processes they use, and their potential to create disaster resilient communities.
5. Understand existing hazard mitigation planning policies and programs.

Key Terms

Administrative capability
Advocacy planning

Capability assessment
Comprehensive plan
Disaster Mitigation Act of 2000
Disaster resilience
Dispute resolution techniques
Emergent groups
Facilitation
Fiscal capability
Focusing events
Hazard analysis
Hazard mitigation
Hazard mitigation committee
Hazard Mitigation Grant Program
Hazard mitigation plan
Hazard mitigation planning
Hazard mitigation policies
Hazard mitigation projects
Hazard mitigation strategy
Hazards management
Land suitability analysis
Land-use planning
Legal capability
Mediation
Multiobjective planning
Negotiation
Physical planning
Plan adoption and implementation
Plan monitoring, evaluation, and modification
Planning process
Police power
Policy dialogue
Policy planning
Political capability
Predisaster mitigation
Public involvement
Robert T. Stafford Disaster Relief and Emergency Assistance Act
Sustainability
Technical capability

Introduction

This chapter will discuss the nexus between hazards analysis and planning, emphasizing disaster resilience and how it fits within the broader concepts of hazards risk management and sustainable development. The hazard mitigation plan provides a tool to link the concepts discussed throughout the text, such as the identification and analysis of hazards, the use of techniques to assess social, economic, and environmental vulnerability (i.e., spatial analysis, modeling, and economic loss estimation), the development of risk management or hazard mitigation strategies, and their application to at-risk individuals, groups, and institutions. The chapter will conclude with a discussion of land-use planning tools and processes and their potential to achieve disaster resilient communities.

Sustainability, Disaster Resilience, and Hazard Mitigation Planning

The concept of sustainability, which has emerged from international nongovernmental organizations, has gained widespread recognition among scholars and practitioners as a sound principle to guide development practices. Sustainable development emphasizes attempts to live in harmony with the natural environment in a manner that provides improved social, environmental, and economic conditions for current and future generations (World Commission on Environment and Development 1987). More recently, the complimentary aims of hazard mitigation (Beatley 1998; Becker 1994; Berke and Beatley 1992; Godschalk et al. 1999; Schwab et al. 1998; Smith and Wenger 2006) and disaster resilience (Beatley 1995; Berke 1995; Burby 2001; Schwab, et al. 1998) have been added to this conceptual framework (see Figure 9.1).*

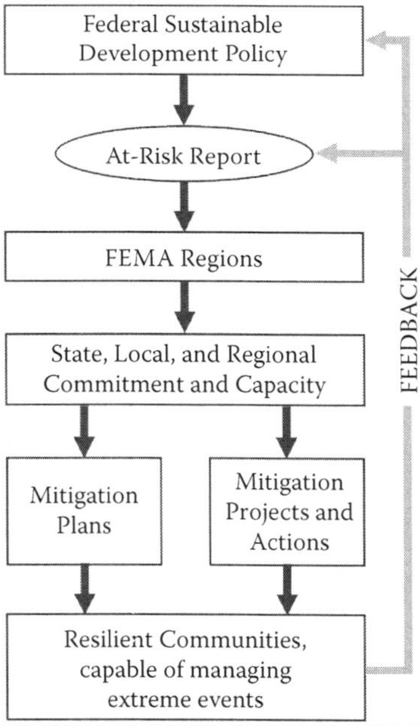

Figure 9.1 A sustainable mitigation policy system (from Godschalk, D. et al. (1999). *Natural Hazard Mitigation: Recasting Disaster Policy and Planning.* Island Press, Washington, DC, p. 531).

A number of international and national commissions and boards have led efforts to link natural hazards mitigation and sus-

* For a summary of hazard scholarship addressing sustainable development themes, see Smith and Wenger (2006: 236).

tainability. The United Nations hosted the Rio Summit, which produced one of the first definitions of sustainability that included hazard mitigation, and declared the 1990s the Decade of Natural Hazard Reduction. The President's Council on Sustainable Development describes specific action items targeting a reduction in governmental subsidies that encourage unsustainable development in known hazard areas (Beatley 1998: 237–238). While a number of international groups and hazards researchers have advocated this position, practitioners at the federal, state, and community level, including land-use planners, have failed to incorporate the concepts of hazard mitigation and disaster resilience into their day-to-day activities on a widespread basis. The ability to link these concepts, not only conceptually, but more importantly through multiobjective planning, can result in mutually reinforcing outcomes and a broader coalition of support across stakeholder groups advocating complimentary positions (Smith and Wenger 2006).

Godschalk et al. (1999) provide a good description of a disaster resilient community and its connectivity to sustainable development principles:

> Resilient communities may bend before the extreme stresses of natural hazards, but they do not break. They are consciously constructed to be strong and flexible rather than brittle and fragile. This means that their lifeline systems of roads, utilities, and other support facilities are designed to continue functioning in the face of rising water, high winds, and shaking ground. It means that their neighborhoods and businesses, their hospitals and public safety centers are located in safe areas rather than in known high-hazard areas. It means that their buildings are constructed or retrofitted to meet building code standards based on the threats of natural hazards faced. It means that their natural environmental protective systems, such as dunes and wetlands, are conserved to protect their hazard mitigation functions as well as their more traditional purposes (p. 526).

As the description suggests, disaster-resilient communities are more sustainable than those that do not develop a comprehensive strategy that incorporates hazard mitigation into their current and ongoing construction, design, and planning activities. Taking appropriate action to ensure greater disaster resilience and sustainability first requires gaining a greater appreciation for the hazards prevalent in the area. Hazards analysis as described in Chapter 1 represents the ongoing and systematic process of identifying and defining the physical (magnitude, scope, and intensity) and temporal (speed of onset, duration) characteristics of hazards, assessing their likelihood of occurrence, and estimating their potential impacts or consequences. Understood in the context of achieving disaster-resilient communities, hazards analysis provides a rational basis for individuals, groups, and organizations to make informed decisions based on that knowledge (see Deyle et al. 1998).

Disaster resilience is achieved through hazards risk management as Shaw describes in Chapter 8. Hazards risk management represents the adoption of a comprehensive and integrative series of practices, policies, and behavior that recognizes how routine and planned actions taken by individuals, groups, and communities affect their level of hazard vulnerability. Decision-making processes are shaped by a number of factors, including access to accurate and timely information, the effectiveness of risk communication and outreach strategies, resource availability, political power and influence, leadership, and the practice of planning. The land-use planning profession offers a meaningful integrative role employing tools, techniques, and processes needed to link the concepts of hazards mitigation, disaster resilience, and sustainable communities.

Increasingly hazard scholars and a growing number of practitioners have recognized that sustainable communities include those that can bounce back from natural hazard events and disasters (Burby 1998; Godschalk et al. 1999; Mileti 1999). The inclusion of hazard mitigation principles in this discourse serves a boundary-spanning function, linking social, economic, and environmental themes to preevent hazard mitigation planning and postevent adjustments to the impacts of disasters (see Figure 9.2). Incorporating hazard mitigation into the routine activities of individuals, governments, businesses, nonprofits, and others represents the ideal manifestation of sustainable development principles. The failure to confront hazards through pre-event planning can cause disastrous consequences as was dramatically evident in New Orleans following Hurricane Katrina. Hazard mitigation planning represents an action-oriented framework used to identify hazards, their expected impact, and measures that can be taken to lessen or eliminate their effects. The

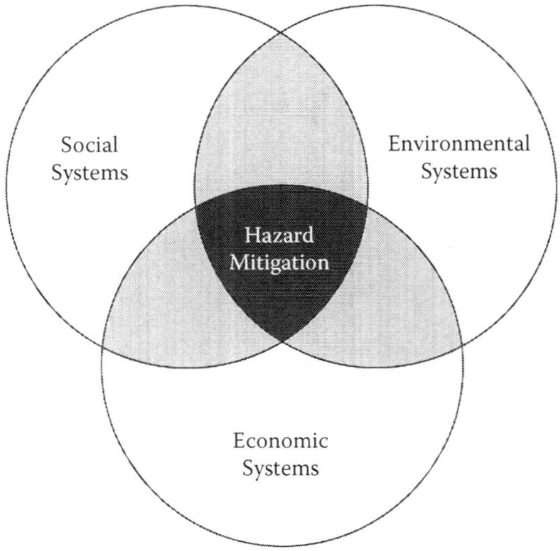

Figure 9.2 Hazard mitigation and sustainable communities.

practice of multiobjective planning provides a procedural vehicle through which complimentary objectives can be achieved before and after disasters.

Hazards and disasters represent a powerful means to understand the preexisting characteristics of communities, as they tend to uncover or highlight social and economic problems that are often tied to issues of race, class, power, equity (see Chapters 4 and 5), and environmental concerns associated with natural system protection and hazard vulnerability (see Chapter 2; Colten 2005). Disasters can serve as a forcing mechanism among communities, causing them to confront problems previously left unaddressed. Examples may include the construction or repair of affordable housing, the incorporation of energy-efficient design principles into new development standards, or changing policies that encourage sprawl into known hazard areas. In the agenda-setting literature, this is referred to as a window of opportunity (Kingdon 1984) or focusing event (Birkland 2006). Major disasters can also provide opportunities to encourage more, not less, development in areas prone to hazards and disasters. Powerful economic interests may use the event, and the ensuing flow of federal assistance, to further a profit-driven agenda, rebuilding communities as quickly as possible, effectively negating the chance to incorporate hazard mitigation or sustainable development principles into the recovery process. The assessment of the political landscape and the impact of competing agendas on the adoption of hazard mitigation strategies should be incorporated into decision-making activities (see Chapter 8), including postdisaster recovery planning.

Critical Thinking: Other than a major disaster, can you think of a focusing event that has occurred in the United States or elsewhere that resulted in significant policy change? Discuss the specific federal, state, and local implications associated with your answer. Is the State of Louisiana's coastal land loss over the past 50 years a good example of a focusing event? Why or why not?

Natural hazards are part of the environmental sphere in which we live. Hurricanes, floods, winter storms, and earthquakes play an important role in the regulation of larger natural systems upon which we all depend. Attempts to physically modify these systems often have severe consequences, including an increased level of hazard vulnerability and damages following disasters. Disasters are a human construct and occur when natural hazards intersect with human settlements or the natural resources upon which people depend. The failure to recognize this reality has resulted in development patterns that are inherently unsustainable (Beatley 1998). Many environmentally sensitive areas such as wetlands, barrier islands, steep-sloped or mountainous areas, and wildlands are prone to natural hazards such as floods, coastal storms, landslides, and wildfire. Limiting development in these areas protects natural areas while reducing the exposure of individuals and communities to the impacts of hazards. For example, wetlands provide a natural buffer against land-falling hurricanes, a reservoir for excess water following floods, a filtration system for pollutants and excess sediment, a recharge area for groundwater, a wildlife habitat, and a site for water-based recreational activities such as canoeing and bird watching.

Conducting a land suitability analysis is regularly used by land-use planners to assess and categorize land according to the type of use that is "most appropriate" based on a series of intrinsic characteristics. Historically this has been done as a means to measure the ecological impact of differing land uses. Planning scholar and practitioner Ian McHarg, in his seminal text, *Design with Nature* (1969), was one of the first to codify this process, incorporating environmental data into the planning process and mapping the results. This method served as the precursor to the development of the geographic information system (GIS) discussed in Chapter 4. The ability to layer and analyze environmental and natural hazards information provides a powerful tool to link land-use decisions to both complimentary geospatial products.* Land suitability analysis may be determined using ecological and natural hazard indicators such as the following:

Topography/elevation/mean sea level (flood, storm surge)
Soil type
  Hydric soils (flood)
  Fill (liquefaction)
  Karst (sinkholes)
Slope (reflecting conditions for avalanche, flash flood, lava flow hazards)
Vegetative type (reflecting conditions for wildfire, mudslide, erosion, or flash flood hazards)
Wetland delineation (to suggest flood vulnerability)
Coastal or riverine erosion (rates)
Existing and future land use
Barrier island presence

This approach also holds promise as a means to capture and visually display the implications of protecting or failing to protect our local, regional, national, and global ecological capital as discussed in Chapter 10.

Critical Thinking: Can you think of other data layers that should be added to the land suitability/natural hazards analysis? Are there other analytical tools that can be used to help conduct a comprehensive hazards analysis? Consider those used in your profession or area of study. Are they currently being used for this purpose? Why or why not? Can this tool be used to assess those natural hazards that are not geographically defined?

---

* In *Design with Nature*, McHarg discusses the connection between design principles that respect coastal ecology and a nor'easter that struck the New Jersey shore in 1962, causing significant damages (see pp. 15–17).

The potential to achieve complimentary benefits must recognize that land prone to natural hazards is often among the most coveted places to live. Examples include the wildland–urban interface, steep-sloped areas, and ocean or riverfront properties. Market demands, reluctance among local officials to limit development in known hazard areas, the inappropriate application of hazard mitigation strategies, and access to large-scale postdisaster assistance programs has encouraged, rather than discouraged this type of development. Further hindering the adoption of a sound hazard mitigation strategy is the fact that the true societal costs associated with development in hazardous areas are not effectively measured, nor are they accounted for in a unified hazards policy. The question of who should bear the costs of living in known high-hazard areas remains one of the most challenging to answer.

Postdisaster assistance policies contribute to this problem. States and local governments tend to view postdisaster assistance as entitlement programs, sought following disasters regardless of pre-event actions taken at the local level that may have increased exposure and vulnerability. The federal government has played an important role in fostering this dependency, as the increasing number of federal disaster declarations are partially the result of political patronage (Platt 1999). It is incumbent on states and local governments to pursue greater local self-reliance relative to the potential impacts of hazards. This concept remains an underemphasized characteristic of sustainable, disaster-resilient communities (Smith and Wenger 2006). The passage of the Disaster Mitigation Act of 2000 was created, in part, to hold states and local governments more accountable and will be discussed next as part of the current hazard mitigation planning policy framework.

Timothy Beatley argues that the practice of hazard mitigation involves a moral dimension that should help frame decision making (1989). Undergirding the conceptual discussion of sustainability is the moral imperative that we should take the actions necessary to ensure the well-being of future generations. The application of ethics to the realm of hazard mitigation requires posing the following questions: (1) How do we reconcile federal, state, and local policies that facilitate rather than hinder choices that increase our vulnerability to hazards? (2) To what extent does local, state, and federal government have a moral obligation to adopt and implement hazard mitigation programs? (3) Does government have a unique obligation to provide additional assistance to the socially vulnerable? (4) To what extent is it the obligation of individuals to take action to reduce their vulnerability to natural hazards versus relying solely on the government for assistance?

Critical Thinking: Do local, state, and federal governments have a moral obligation to protect life and property from the impacts of natural hazards? If so, how do you reconcile the fact that existing policies have the unintended effect of encouraging, rather than discouraging development in known hazard areas? What would you do to address this dilemma? Development in known high-hazard areas is often comprised of secondary or vacation homes. To what extent is it the obligation of government to provide financial assistance to protect these properties?

The Hazard Mitigation Planning Policy Framework

The Disaster Mitigation Act of 2000 requires that state and local governments must develop hazard mitigation plans in order to remain eligible for pre- and postdisaster hazard mitigation funding, including the Pre-Disaster Mitigation (PDM) and the Hazard Mitigation Grant Program (HMGP). The Disaster Mitigation Act further codifies the federal rules and requirements associated with the development of state and local hazard mitigation plans. Prior to that time, states were required to develop plans as stipulated by the Robert T. Stafford Disaster Relief and Emergency Assistance Act. The Stafford Act, which was passed by Congress in 1988, created three key disaster recovery programs: the Individual Assistance, Public Assistance, and Hazard Mitigation Grant Programs.* The Stafford Act emphasizes the administration of these programs rather than a broad policy framework advancing the concepts of hazard mitigation and planning for disasters (Godschalk et al. 1999; Mileti 1999). Godschalk and his colleagues found in their study of plans predating the Disaster Mitigation Act that the quality of state hazard mitigation plans was weak and their ability to foster the implementation of a comprehensive hazard mitigation strategy was limited (1999). A more recent study conducted by the Government Accounting Office cited the failure of the federal government to develop a national policy framework to provide guidance on the use of numerous, but disjointed, hazard mitigation techniques to more effectively reduce future disaster losses (Government Accountability Office 2007).

Hazard mitigation plans should provide a framework for action, in the form of a series of interrelated programs, policies, and projects designed to reduce the level of hazard vulnerability in a given area. Most local hazard mitigation plans have been created as a means to an end, namely access to hazard mitigation grant program funding (that addresses problems created in the past) rather than a future orientation that seeks to guide development and human settlement patterns in a manner that reflects hazard risk and vulnerability. The failure to adopt policies that reflect this future orientation is not sustainable, nor disaster resilient. The hazard mitigation planning process is described next as a means to address these aims. The successful use of this tool can be greatly improved if tested planning techniques and processes are effectively utilized.

* The Individual Assistance program provides grants and loans to assist homeowners and renters make repairs to damaged homes, while the Public Assistance program provides federal funding to assist states, communities, and nonprofits with offsetting the costs of disaster response efforts, cleaning up disaster generated debris, and repairing damaged public infrastructure.

The Hazard Mitigation Plan

The hazard mitigation plan is comprised of several parts: the hazard mitigation planning committee, hazard identification and analysis, vulnerability assessment, capability assessment, hazard mitigation strategy, and plan adoption and implementation. The hazard mitigation committee is tasked with the creation and implementation of the hazard mitigation plan. Community hazard mitigation committee membership might include:

Public works director
Land-use planner
Local floodplain administrator
Local emergency manager
Building official/building inspector
City manager/assistant city manager
Finance officer or director
Economic development director
County representative
Citizen group representatives
Nonprofit representatives
Local business representatives
Utility company representative

The committee may establish subcommittees based on key issues identified over time. Examples may include functional topics associated with vulnerable housing and infrastructure or specific challenges associated with future growth or desired changes in existing land-use regulations.

It is the responsibility of the committee to solicit public involvement throughout the planning process. The public, for example, should play a role in the identification of hazards and their potential impact. This provides for the inclusion of local experiential knowledge that can be missed by local officials or outside consultants hired to help develop the plan. Examples may include anecdotal information based on past experiences with hazards and disasters, or the accumulation of newspaper clippings and photographs taken by individuals. Public input also provides a vehicle to obtain a collective understanding of local risk perception, which should be used to frame education and outreach strategies (see Chapter 7). The power of participatory planning and the influence of the planning process will be discussed later in this chapter.

Critical Thinking: Can you think of others that should be included in a hazard mitigation committee? Should participation or membership change over time? What role might individuals from your area of study or profession play on the hazard mitigation planning committee?

Following the creation of a committee, members should begin the background studies (hazard identification and analysis, vulnerability assessment, and capability assessment) that serve as the factual basis for the hazard mitigation plan. In order to develop a sound hazard mitigation strategy, it is necessary to identify and analyze the hazards present in the study area (see Table 1.1 for an overview of natural hazards prevalent in the United States). Once the hazards are identified, they are analyzed by documenting their historic occurrence and associated impacts. Newspapers, weather service reports, technical documents, and personal accounts are representative of the types of information sources typically used for this purpose. The historical documentation of past events provides some insight into what an area could expect in the future. The review of past events does not provide a good means to estimate the probability of future disasters. This requires conducting a vulnerability assessment. The vulnerability assessment involves the evaluation of hazard risk, including the likelihood that events of a varied magnitude will impact the study area, causing a series of estimated damages. As discussed in Chapter 4, some segments of a population are more vulnerable than others due to social and economic characteristics. A sound vulnerability assessment estimates future expected losses based on projected development patterns and demographic changes over time.

As noted in Chapter 7 and this chapter, respectively, the creation of hazards base maps is a useful means to identify and assess geographically defined hazards. The delineation of hazards using GIS allows for the georeferenced analysis of hazards relative to the built and natural environment. A GIS platform, coupled with the tools discussed in Chapter 3, enables the modeling of various hazard scenarios, providing valuable "what if" information that should guide the development of the hazard mitigation strategy that follows. The ability to quantify expected losses as described in Chapter 5 provides an important means to establish a rational basis for action discussed in Chapter 8 and communicate risk (see Chapter 7). The real power of this type of assessment is the ability to estimate how the level of vulnerability can be increased or decreased by taking specific actions. For example, continued development in known hazard areas or the adoption of more stringent building codes can be modeled, and their negative and positive impacts on hazard exposure and vulnerability measured.

The vulnerability of a community is partly a function of the existing technical, fiscal, administrative, legal, and political capabilities of a jurisdiction to reduce, manage, or eliminate the effects of identified hazards. Elements and indicators of local capability include:

Technical (GIS, visualization, hazard modeling software, planning)
Fiscal (local budget, grants-in-aid, loans, capital investments)
Administrative (technical and administrative staff, contractors)
Legal (existing rules and regulations, legislation)
Political will (votes cast, policies and regulations adopted, land-use decisions)

Analyzing this capability requires the review of existing documents, plans, programs, and policies. Local documents may include the local comprehensive plan, local flood damage prevention ordinance, and capital improvements plan, among others. Local policies that target other aspects of a community (i.e., economic development, environment protection, or social services) may impact the vulnerability of a community to natural hazards. The capabilities of additional stakeholders, such as nonprofits and the private sector, should be included in this assessment. Other examples may include federal and state regulations, regional agreements, or the multijurisdictional sharing of resources. A comprehensive capability assessment revolves around two basic questions: (1) does the community posses the tools needed to confront their vulnerability to hazards, and (2) do existing policies, plans, programs or activities currently in place increase or decrease current or future hazard vulnerability?

Technical capability refers to the access to and use of analytical tools, including GIS, visualization, and hazard-modeling software. The fiscal capability of a jurisdiction includes their internal and external access to financial resources and the commitment of these resources to hazard mitigation–related activities. Internal resources include the regular operating budgets of participating stakeholder groups, while external resources include funds obtained through the procurement of grants and loans. Administrative capability refers to the staff, personnel, or contractors available to create, monitor, and implement the hazard mitigation plan and associated strategies.* Legal capability refers to the type and strength of government rules and regulations that provide the legal standing to act. Federal, state, and local government agencies are responsible for enforcing regulations and policy. The adoption and enforcement of most land-use planning techniques are done at the local level.† Local governments possess the legal authority, or what is referred to as "police power," to enact land-use planning measures necessary to protect the health, safety, and general welfare (Nolon and Salkin 2006). Less well recognized is the application of this concept to protect citizens and their property from the impacts of hazards and disasters.

The adoption of local hazard mitigation strategies are often driven by the level of political will present in a community. Political capability is the willingness of elected officials and parties, such as developers and business leaders, to support the adoption of hazard mitigation strategies. It also refers to the level of acceptance among citizens and community groups, including those who may advocate

* Following disasters, the ability to implement hazard mitigation strategies can be compromised, as staff are often overwhelmed with the management of postdisaster aid programs. The postdisaster environment can also provide an opportunity to hire additional staff through grants triggered by a federal declaration. The ability to retain staff and their highly valued experienced-based knowledge is often difficult, as these skills are in high demand among federal, state, and local government agencies as well as private sector consultants and contractors.

† The common refrain among practicing planners, "all planning is local," is not entirely accurate. In reality, federal and state policies can significantly impact land use at the local level.

stronger regulations. Adopting policies restricting land-use activities in identified hazard areas or more stringent building codes and ordinances are representative of actions that require a high level of political will. Convincing skeptics who believe that such measures unnecessarily restrict market choices or land use, or that the federal government will pay for the costs of reconstruction following a disaster, requires developing a sound counterargument backed up with verifiable data and political support. It also requires moving beyond the realities of short election cycles on which officials tend to base many of their policy decisions. Effective strategies engage a broad coalition of elected and nonelected individuals in positions of local power or influence and describe the tangible economic, environmental, and social benefits of hazard mitigation that resonate with their own agenda in order to achieve and sustain changes in the status quo.

In addition to the assessment of technical, fiscal, administrative, legal, and political capabilities, the hazard mitigation plan should identify gaps in existing tools used to confront hazard vulnerability. Identified gaps can be translated into the adoption of policies addressing these issues. For example, during the assessment of existing policies it may be found that the community is a participating member of the National Flood Insurance Program, but not the Community Rating System (CRS), which requires the adoption of a more comprehensive approach to floodplain management. Based in part on this review of existing policies, the community may choose to join the CRS as a part of their hazard mitigation strategy.

Critical Thinking: Is one element of the capability assessment more important than another? If so, explain why.

The hazard mitigation strategy represents a series of policies, programs, and projects chosen to reflect the actions necessary to address the results of the vulnerability and capability assessments. A vulnerability assessment identifies at-risk structures and populations as well as future vulnerabilities based on projected growth. This information is used to develop specific strategies designed to reduce identified vulnerabilities (Figure 9.3). An effective hazard mitigation strategy requires the use of multiple hazard mitigation projects and policies identified by a diverse team of stakeholders. Potential hazard mitigation projects may include the relocation or elevation of flood-prone properties, the retrofitting of vulnerable community facilities, and the hardening of public infrastructure. Hazard mitigation policies may include joining the National Flood Insurance Program, limiting the placement of public infrastructure and critical facilities in identified hazard areas, or adopting building code standards reflecting the latest findings of the hazards analysis.

As techniques are selected, appropriate funding or administrative support is identified. In some cases, the adoption of mitigation techniques will require amending existing regulations or creating new ones. It is therefore incumbent on the hazard mitigation committee to involve elected officials early in the process in order to gain their support. It is also important to make sure that the mitigation techniques chosen can be accomplished, given existing capabilities or government officials

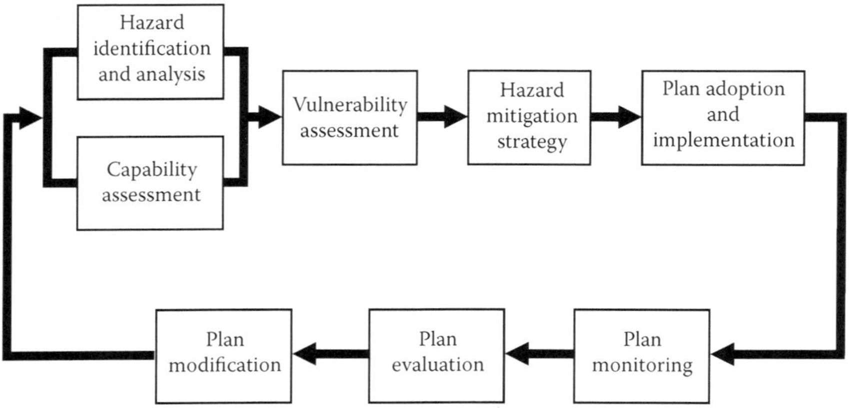

Figure 9.3 The hazard mitigation planning process.

committed to procure the resources needed to accomplish them. Otherwise, their implementation will be compromised. The following examples of hazard mitigation techniques were adapted from *Keeping Hazards from Becoming Disasters: A Mitigation Planning Guidebook for Local Government,* prepared by North Carolina Division of Emergency Management in 2001.

Prevention (planning, zoning, floodplain, and subdivision regulations; open space preservation, storm-water management, drainage system maintenance, capital improvements programming, shoreline/fault zone setbacks)

Property protection (relocation, acquisition, elevation, critical facilities protection, insurance, retrofitting of hazard-prone structures)

Natural resource protection (floodplain protection, beach and dune preservation, riparian buffers, conservation easements, erosion and sediment control, wetland preservation, slope stabilization)

Structural projects (reservoirs, seawalls, levees, channel modifications, beach nourishment)

Public information (outreach projects, hazard map information, real estate hazard disclosure, warning systems, library, hazards expo)

Critical Thinking: How would you balance the selection of hazard mitigation projects and policies based on the findings of the risk assessment with the level of local capability to implement them? Which hazard techniques do you believe are the most effective? Why?

Plan adoption and implementation is not the final step, as the planning process is more accurately characterized as a continually revised series of actions (Figure 9.3). Once written, the adoption of the plan gives it legal standing among those responsible for its implementation and the larger community as a whole. In order for the plan to be an action-oriented document, it should contain a strong implementation strategy. It is critically important that the plan is routinely monitored to make sure that those individuals, agencies, and organizations assigned various tasks are completing them in accordance with the plan schedule. The efficacy of the plan requires holding participants accountable. If plan goals and objectives are not being met, the planning committee should take the appropriate corrective actions. An effective plan monitoring, evaluation, and modification strategy includes the means to integrate the latest understanding of hazard vulnerability and capability, as both change over time. New development, changes in the makeup of the population, or the use of more refined analytical tools influence our understanding of hazard vulnerability. Similarly, the capability to mitigate hazards can change. The formulation of new or modified policies should be incorporated into the document. Finally, it is important to monitor progress. Attaining mitigation goals should be documented and their benefits measured, when possible (see Chapter 8). If monetary benefits can be ascertained (i.e., losses avoided due to pre-event actions) this information can garner continued political support, among both local elected officials that may choose to adopt more progressive mitigation policies and federal agencies that may be more willing to provide pre- and postdisaster hazard mitigation funding or other types of assistance.

Critical Thinking: What indicators provide a basis to assess the political will of a community to adopt a comprehensive hazard mitigation strategy? Once identified, how would your mitigation strategy reflect that reality and account for changes associated with election cycles and turnover in personnel? What role do citizens play in shaping the political will of a community?

The Power of Plan Making: Tools and Process

Land-use planning is one of the most powerful tools to advance the aims of hazard mitigation and disaster resilience (Burby 1998). A review of land-use planning can be understood within a process-oriented framework tied to physical and policy-related outcomes associated with the location, type, and density of development (Chapin and Kaiser 1985). Physical planning traces its roots to the development of zoning and the separation of land uses, whereas policy planning is linked to the traditions of decision sciences and the more recently emerging field of policy analysis (Friedmann 1973). Both rely on the other through the collection, analysis, and display of spatial information used to produce a series of recommended actions, including policies and physical plans.

Land-use planners use a series of land-use planning tools that are directly relevant to hazards risk management (Berke and Beatley 1992; Burby 1998; Godschalk et al. 1989). Most communities have developed a number of planning documents that are intended to guide their actions. Local plans may include the comprehensive plan, the capital improvements plan, and area plans (central business district plan, and parks and recreational plans). Communities may also participate in the creation of multijurisdictional plans including growth management plans, watershed management plans, and transportation plans. Comprehensive plans provide broad policy guidance regarding future growth.* The ability to guide development in a manner that respects identified hazard areas is an important long-term aspect of hazard mitigation and should be included in a community comprehensive plan. The capital improvements plan provides an estimated timeline for future capital development projects such as water, sewer, roads, and schools and the means to finance their construction. The placement of public infrastructure plays a crucial role in shaping development patterns. The degree to which the identification and analysis of natural hazards is used in the "formulation, design or justification of land use and management tools" (Godschalk et al. 1998) varies widely, while conducting a risk analysis as precondition of their use is rarely employed, as Table 9.1 suggests.

The Planning Process: Building Stakeholder Capacity to Confront Hazards

The ability to effectively convey the implications of the hazards analysis to decision makers and the general public and facilitate the conditions under which appropriate action is taken requires a process-oriented approach. The plan is often superseded in importance by the planning process undertaken to create it (Innes 1995, 2004). The process of collaborative plan making facilitates locally driven decisions and serves as an important communication venue, educational forum, and means to build a coalition of supportive stakeholders. Public participation, consensus-building, and process-oriented learning has a rich history in the field of planning. In Sherry Arnstein's classic *Eight Rungs on the Ladder of Citizen Participation* (1969), she describes several levels of participation and their implications (Figure 9.4). Manipulation and therapy reflect efforts by those in positions of power to "educate" or "cure" participants rather than allow them to voice their opinion. Informing and consultation are characterized by actions that allow citizens to provide input, but the information is rarely acted upon or implemented. Placation allows for citizens to

* Planning naturally involves the act of plan making. Among the most widely recognized community planning documents is the comprehensive plan (see Berke et al. 2006; Kent 1964). The comprehensive plan is comprised of five defining characteristics: (1) a guide for the physical growth and development of a jurisdiction or area; (2) a long-range orientation based on a broad vision, including the steps necessary to achieve it; (3) comprehensive in nature, addressing city functions (i.e., transportation, housing, public infrastructure and facilities, land use, and parks and recreation) and their interconnectivity; (4) a set of coherent policies intended to shape the quality, quantity, type, and density of development; and (5) a tool used by the local governing body to guide decision making (So and Getzels 1988: 60–61).

Table 9.1 Use of Hazard Assessment in Land-Use Planning

Plans and Implementation Tools	*Hazard Identification*	*Vulnerability Assessment*	*Risk Analysis*
Planning			
Comprehensive plan, hazard component of comprehensive plan, recover/reconstruction plan	CP	S	R
Local emergency management plans	CP	CP	R
Development Regulations			
Zoning ordinance	S	R	R
Subdivision ordinance	S	R	R
Hazard setback ordinance	CP	R	R
Building Standards			
Building code	CP	R	R
Special hazard resistance standards	CP	S	R
Retrofit standards for existing buildings	CP	S	R
Property Acquisition			
Acquisition of undeveloped lands	CP	R	R
Acquisition of development rights	CP	R	R
Building relocation	CP	CP	R
Acquisition of damaged buildings	CP	CP	R
Critical and Public Facilities Policies			
Capital improvements programs	CP	R	R
Location requirements for critical facilities	CP	S	R
Location of public facilities and infrastructure in less hazardous areas	CP	R	R

—continued

Table 9.1 Use of Hazard Assessment in Land-Use Planning (continued)

Plans and Implementation Tools	*Hazard Identification*	*Vulnerability Assessment*	*Risk Analysis*
Taxation			
Impact taxes	CP	S	R
Reduced or below-market taxation	CP	R	R
Information Dissemination			
Public information program	CP	S	R
Hazard disclosure requirements	CP	S	R

Note. *CP = common practice; S = sometimes used; R = rarely, if ever, used.*

Source: Deyle, R., S. French, R. Olshansky, and R. Patterson (1998). In Cooperating with Nature: Confronting Land-Use Planning for Sustainable Communities, ed. R. Burby. Joseph Henry Press, Washington, DC, p. 123.

"advise," but the power to act remains with others in positions of authority. Citizen power includes partnership, delegated power, and citizen control. Partnership allows for negotiated agreements and associated "trade-offs," while delegated power and citizen control involve the transfer of decision-making authority. These categories are particularly useful to frame the decision-making process described in Chapters 7 and 8 and the role of dispute resolution techniques that are discussed next.

Critical Thinking: Apply the Eight Rungs on the Ladder of Citizen Participation to the concepts discussed in Chapter 7, Risk Communication. What changes would you recommend to existing risk communication and decision-making strategies based on Arnstein's framework?

Like Arnstein, Paul Davidoff argues that urban planners should move away from their strict belief in the role of the neutral technician, instead embracing the concept of advocacy planning. This approach is intended to provide a voice for the disenfranchised, more socially vulnerable populations described in Chapter 4. Alinsky (1969) notes that planning should be driven by community organizing, a concept often applied postdisaster through emergent groups that form spontaneously based on an identified need that appears to be unmet by existing agencies, organizations, or groups (Drabek 1986).

Pre- and postdisaster issues surrounding hazard mitigation are often contentious, particularly when actions are based on governmental mandates rather than collaboratively derived approaches. Examples may include the adoption of building codes, limiting development in identified hazard areas, or other mitigation

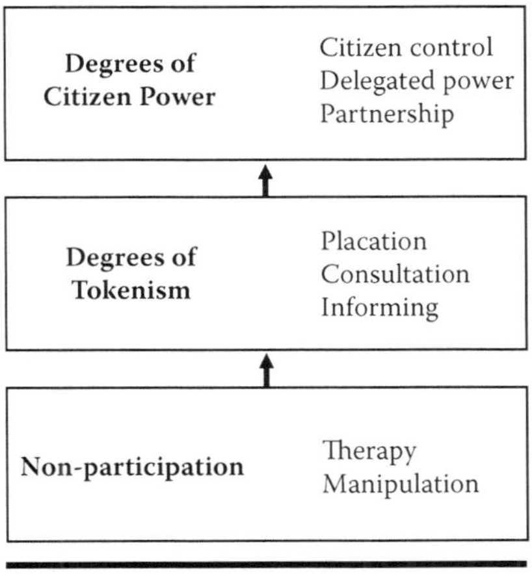

Figure 9.4 Ladder of citizen participation.

strategies that place restrictions on the location, type, or density of development. These actions often fall within the purview of local land-use planners, and they should play an important role in the resolution of these conflicts. Following disasters, a tension exists between speeding the reconstruction process and taking a more deliberative approach (Haas et al. 1977). Development interests and individual citizens place a great deal of pressure on elected officials and their staff to streamline policies and planning processes in order to regain a sense of normalcy. The postdisaster environment can also provide a unique window of opportunity to embrace the tenets of sustainable redevelopment, including the integration of hazard mitigation concepts into recovery policy and the physical reconstruction of an impacted area (Smith and Wenger 2006). The adoption of hazard mitigation measures in the predisaster environment enables communities to bounce back following disaster (Beatley 1998). Achieving a disaster-resilient and a more sustainable community is not accomplished by a single act, or a series of actions, but rather an on-going process.

Understood in the context of hazards and disasters, the planning process necessarily involves (1) a continual assessment and reassessment of hazards, including their likelihood of occurrence and expected impacts based on new, more accurate hazard information; (2) the adoption of new polices, development choices, and settlement patterns that increase or decrease hazard vulnerability; and (3) the creation and maintenance of a participatory decision-making framework based on the best available information regarding hazards, the meaningful involvement of stakeholders, and the adaptation of participatory methods based on local political, social, and institutional conditions. A number of proven dispute resolution techniques can be used to advance what Arnstein refers to as "citizen power." Traditional ad hoc participatory approaches (e.g., public hearings) often fail to advance egalitarian objectives (Kemp 1985) or provide a sound means to communicate hazard risk (Kasperson 1986). The power of dispute resolution lies in its ability to seek consensus among competing groups, including those historically excluded from decision making. Care must be taken by those employing these techniques to ensure that the approach does not become a tool used by technical experts to advance the status quo advocated by those in power, which Arnstein refers to as "tokenism" (1969). Planners have employed these techniques with increasing regularity and represent one of several groups, including diplomats, organized labor, and attorneys, that consider these skills a part of their professional training.

Dispute resolution techniques are representative of tools that are applicable to multistakeholder dilemmas, including those associated with hazard mitigation (Godschalk et al. 1998). Specific methods include facilitation, negotiation mediation, and policy dialogue. The role of the facilitator is to provide the best possible conditions to foster effective communication between parties. This may include establishing clear rules and conditions, ensuring appropriate stakeholder representation, keeping the group focused on the issues rather than personalities, and asking questions of participants in order to expose possible areas of agreement. State officials may assume this role, because they are often required to bridge policy debates between federal programs and their impact on local communities. Negotiation is a bargaining process conducted between two or more parties, each seeking a satisfactory agreement. In practice, negotiation can lead to suboptimal agreements defined by the zero-sum game principle in which the benefit one party derives is necessarily subtracted from the other. Thus the challenge becomes, under what conditions can negotiations lead to mutually beneficial agreements or a non-zero-sum game (see Raiffa 1982)? Crafting desired changes in federal hazards policy is often the result of local and state officials who possess effective negotiation skills (Smith 2004). This may include, for example, the development of postdisaster hazard mitigation policies that reflect local needs.

Disputes among multiple parties can prove difficult, given the multitude of interests and their differential access to information and power. Mediation involves the use of a trained, neutral third party who attempts to assist conflicting parties reach a mutually beneficial agreement, recognizing and neutralizing power imbalances (i.e., "leveling the playing field") and emphasizing the sharing of information.* While the use of mediation has not been widely adopted as a means to resolve disputes associated with hazards, it offers promise and merits further study.

The procedural nature of policy making routinely includes the use of policy dialogue. This technique emphasizes ongoing discussions in which participants gain a greater understanding of the other's positions, but more importantly their interests, and use this knowledge to seek resolution to policy dilemmas. Iterative communication among stakeholders can elicit cooperation (Axelrod 1984). In those cases where disputes are not resolved, participants often claim to have gained valuable insights into differing points of view while improving communication channels (Bingham 1986). Continued interaction can also help identify unique and far-reaching decisions that may remain otherwise hidden. As knowledge among stakeholders grows and is shared with others, collectively derived solutions become possible. As noted in the earlier discussion of the capability assessment, a participatory process should lead to the development of realistic policies and procedures and identify gaps that merit attention. The Charlotte/Mecklenburg case study clearly shows that the involvement of stakeholders in the decision-making process

* Mediated agreements are not legally binding. This differs from arbitration, which results in a legally binding decision reached by a third party such as a judge or professional arbitrator.

increases the likelihood of adopting hazard mitigation measures based on information derived from a participatory process.

Critical Thinking: Based on personal experience or observation, provide a positive and negative example of the public participation process. If you were involved in the example discussed, did you feel like your participation or the participation of others made a difference? Why or why not? Did the process employ dispute resolution techniques? Were they effective?

Meaningful policy shifts can occur, given the appropriate circumstances. Examples may include major disasters that trigger federal recovery funding or local demands to address immediate problems. In 1995, the remnants of Hurricane Danny caused significant flooding in the city of Charlotte and Mecklenburg County, North Carolina, but it did not result in a presidential disaster declaration. This caused the city and county to investigate local measures to reduce flood losses while maintaining their ongoing efforts to improve water quality and preserve the natural functions of the floodplain without unnecessarily limiting growth. One public official saw the process as a way to avoid what he termed the "hydro-illogic" cycle—a process characterized by the initiation of a postevent stormwater planning study recommending an expensive means to address the problem. Elected officials did not act on the proposed measures, leaving the city vulnerable to future flooding.

Several factors lead to a successful resolution of this policy dilemma. First, the county sought to create a vision that reflected the identified characteristics of the area. Charlotte is a major southern city that prides itself on being a place that welcomes growth. It is also prone to serious flooding. A broad-based collection of interests was brought to the table to craft an acceptable solution. Participants included developers, environmentalists, community organizations, planners, engineers, county commissioners, city officials, and their staff. Over two months were spent identifying and clearly defining the problem rather than prematurely discussing possible solutions. Initially, many grew impatient. In hindsight, coalition members realized that the work done up front set the stage for success. Those involved in the policy dialogue sought to resolve identified problems (reducing flood losses and improving water quality) rather than dwell on the symptoms, perceptions, or emotions. For example, developers noted that they did not want to build homes that flood, nor did they want to be blamed for the construction of flood-prone homes in the 1960s. This type of honest dialogue provided room for discussion, and the realization that access to information was the most pressing need to drive meaningful policy change. Efforts were made to stay focused rather that straying too far from the agreed-upon problem, thereby avoiding "deal killers" (interview, Canaan 2001).

The end result was the creation and adoption of a city/county guidance document that identified six primary strategies, including new floodplain development standards, an enhanced flood warning system, a drainage system maintenance plan, a public information campaign, an interagency steering committee, and the development and implementation of watershed-based hazard mitigation plans. Each strategy was tied to specific action-oriented tasks with associated timelines and identified parties responsible for their implementation. An overriding theme of the guidance document was the reliance on verifiable data to guide policy choices.

The basis for policy formulation was the creation of new Flood Insurance Rate Maps or FIRMs. The city and county chose to create a map that was not a static depiction of current conditions, like those typically created by FEMA.* The maps were designed to reflect the future flood hazard conditions present once all allowable development had occurred in the floodplain. This was accomplished by reviewing existing zoning maps and estimating future settlement patterns and land uses. This approach reflected a significant departure from traditional FIRMs, since new development can cause changes in flood elevations, given the increase in impervious surfaces and the placement of fill material in the floodplain.

The Charlotte/Mecklenburg Stormwater Services group used a number of scenarios that clearly demonstrated the relationship between development and future flood risk. Not only did the team show how flood elevations change given differing floodplain build-out scenarios, they also quantified expected flood losses under these conditions. By tying proposed actions or inaction to real dollar figures, developers and county commissioners took notice. Developers asked for repeated iterations of differing scenarios, ultimately recognizing that mapping the flood hazard as it would look in the future and regulating to an agreed-upon standard based on this information, made sense. The "future conditions" mapping effort not only increased flood elevations by as much as eleven feet, it also expanded the floodway, the area within the floodplain where development is severely restricted.

Prior to Hurricane Danny, the city and county sought to alleviate significant water quality problems. The county commissioners established a goal of improving water quality to a level that would allow for "prolonged human contact." In order to achieve this aim, the commission and the county's Department of Environmental Protection established the Surface Water Improvement and Management Panel (SWIM). The coalition of environmentalists, citizens, developers, and local officials created a stream buffer plan that identified buffer widths based on the acreage drained by each creek or stream. The larger the drainage area, the larger the buffer required. If the buffer area exceeded the mapped floodplain, new development was not allowed within the larger boundary. By keeping the area along the creek free of development, the existing vegetation filtered pollutants, while the open space provided for additional water storage. By using information that the developers recognized as valid, county and city officials were able to negotiate an agreed-upon regulatory framework.

* The City of Charlotte and Mecklenburg County negotiated an agreement with FEMA to develop the FIRMs. The formal agreement designated the city and county a "Cooperating Technical Partnership" which enabled them to use the maps to regulate floodplain development and established defined roles for all parties.

The new maps served as the basis for the city and county's flood hazard analysis component of their watershed-based hazard mitigation plans. The resulting hazard mitigation strategy emphasized the relocation of flood-prone homes and a number of other complimentary objectives. Once specific homes were identified, the county sought funding to remove the structures from the floodplain. Mecklenburg County received approximately 10 million dollars in HMGP funds to acquire 116 homes. County storm water fees provided an additional 2.2 million dollars. The acquired properties were demolished, and the land reverted to open space. The open space was used to expand an existing greenway and the stream buffer system, thereby reducing flood risk, providing local recreational opportunities, and improving water quality, all key objectives identified in the early stages of the consensus-building process and representative of the nexus between disaster resilience and sustainable development.

The final chapter describes a proposed hazards management policy framework that links the concepts discussed throughout the text and suggests specific modifications to existing federal hazard mitigation policies and programs that will facilitate the creation of disaster-resilient and, hence, more sustainable communities. The hazards management policy framework emphasizes several important elements, including a greater emphasis on the connection between land-use planning and hazards, assessing current and future hazard vulnerability, building local capacity and self reliance, balancing policy incentives and penalties, and measuring success.

Discussion Questions

Should a community hazard mitigation plan be incorporated into a community's comprehensive master plan or serve as a stand-alone planning document? Why or why not?

This chapter proposes to use land-use planning as a means to integrate hazard mitigation, disaster resilience, and sustainable development principles. Are there other approaches or disciplines that should be considered? If so, what are they, and how would you suggest they function?

Discuss the act of plan making and the planning process in the context of reducing hazard vulnerability. Is one necessarily more important than the other?

Applications

You Be the Planner

What planning tools and processes would you employ to address the results of a community hazards analysis and why? Are some tools hazard specific? Are your choices based on local capability? How would you facilitate coalition building and the involvement of multiple stakeholders?

Web Sites

American Planning Association. http://planning.org/
Association of State Floodplain Managers. http://www.floods.org
Community Rating System. http://www.fema.gov/business/nfip/crs.shtm
Federal Emergency Management Agency. http://www.fema.gov/
Federal Emergency Management Agency, Higher Education Project. http://training.fema.gov/EMIWeb/edu/
Institute for Business and Home Safety. DisasterSafety.org. http://www.disastersafety.org/
National Emergency Management Association. http://www.nemaweb.org/
National Oceanic and Atmospheric Administration. http://www.noaa.gov
National Oceanic and Atmospheric Administration Coastal Services Center. NOAA CSC Products and Services: Coastal Resilience. http://www.csc.noaa.gov/bins/resilience.html
National Response Framework Resource Center. http://www.fema.gov/NRF
Public Entity Risk Institute. http://riskinstitute.org/

References

Alinsky, S. (1969). *Reveille for Radicals.* Vintage Books, New York.
Arnstein, S. R. (1969). A ladder of citizen participation. *Journal of the American Institute of Planners* 35:216–224.
Axelrod, R.(1984). *The Evolution of Cooperation.* Basic Books, New York.
Beatley, T. (1989). Towards a moral philosophy of natural disaster mitigation. *International Journal of Mass Emergencies and Disasters* 7(1):5–32.
Beatley, T. (1995). *Promoting Sustainable Land Use: Mitigating Natural Hazards through Land Use Planning.* Hazard Reduction and Recovery Center Publication. No 133A. Texas A&M University, College Station, TX.
Beatley, T. (1998). The vision of sustainable communities in Cooperating with Nature: Confronting Natural Hazards with Land-Use Planning. R. Burby. John Henry Press, Washington, DC, 233–262.
Berke, P. (1995). Natural hazard reduction and sustainable development: a global assessment. *Journal of Planning Literature* 9(4):370–382.
Berke, P. and T. Beatley (1992). *Planning for Earthquakes: Risk, Policy and Politics.* Johns Hopkins University Press, Baltimore, MD.

Berke, P., D. Godschalk, E. Kaiser, and D. Rodriguez (2006). *Urban Land Use Planning*, 5th Edition. University of Illinois Press, Champaign, IL.

Bingham, G. (1986). *Resolving Environmental Disputes.* Conservation Foundation, Washington, DC.

Birkland, T. A. (2006). *Lessons of Disaster: Policy Change after Catastrophic Events.* Georgetown University Press, Washington, DC.

Burby, R. (1998). *Cooperating with Nature: Confronting Natural Hazards with Land-Use Planning for Sustainable Communities.* Joseph Henry Press, Washington, DC.

Burby, R. (2001). *Building Disaster Resilient and Sustainable Communities.* Federal Emergency Management Agency, Emergency Management Institute, Higher Education Project. The course is available online at http://training.fema.gov/emiweb/edu/completeCourses.asp.

Burby, R., S. French, B. Cigler, E. Kaiser, D. Moreau, and B. Stiftel (1985). *Flood Plain Land Use Management: A National Assessment.* Westview Press, Boulder, CO.

Canaan, D. (2001). Director, Mecklenburg County Stormwater Services. Interview.

Chapin, F. S. and E. J. Kaiser (1985). *Urban Land Use Planning*, 3rd Edition. University of Illinois Press, Urbana, IL.

Colten, C. (2005). *An Unnatural Metropolis: Wresting New Orleans from Nature.* Louisiana State University Press, Baton Rouge.

Deyle, R., S. French, R. Olshansky, and R. Patterson (1998). Hazard assessment: the factual basis for planning and mitigation. In *Cooperating with Nature: Confronting Land-Use Planning for Sustainable Communities*, ed. R. Burby. Joseph Henry Press, Washington, DC, pp. 119–166.

Drabek, T. (1986). *Human System Responses to Disaster: An Inventory of Sociological Findings.* Springer-Verlag, New York.

Federal Emergency Management Agency (1995). *National Mitigation Strategy.* January. FEMA, Washington, DC.

Friedmann, J. (1973). *Retracking America: A Theory of Transactive Planning.* Anchor Press, Garden City, NY.

Godschalk, D., D. Brower, and T. Beatley (1989). *Catastrophic Coastal Storms: Hazard Mitigation and Development Management.* Duke University Press, Durham, NC.

Godschalk, D., E. Kaiser, and P. Berke (1998). Integrating hazard mitigation and land use planning. *In Cooperating with Nature: Confronting Natural Hazards with Land-Use Planning for Sustainable Communities*, ed. R. J. Burby. Joseph Henry Press, Washington, DC, pp. 85–118.

Godschalk, D., T. Beatley, P. Berke, D. Brower, and E. Kaiser (1999). *Natural Hazard Mitigation: Recasting Disaster Policy and Planning.* Island Press, Washington, DC.

Government Accountability Office (2007). *Report to the Ranking Member, Committee on Financial Services, House of Representatives. Natural Hazard Mitigation: Various Mitigation Efforts Exist, but Federal Efforts do not Provide a Comprehensive Strategic Framework.* Washington, DC: United States Government Accountability Office. GA0–07–403.

Haas, J., R. Kates, and M. Bowden (1977). *Reconstruction Following Disaster.* MIT Press, Cambridge, MA.

Innes, J. (1995). Planning theory's emerging paradigm: communicative action and interactive practice. *Journal of Planning Education and Research* 14:183–189.

Innes, J. (2004). Consensus building: clarification for the critics. *Planning Theory* 3(1):5–20.

Kasperson, R. (1986). Six propositions on public participation and their relevance for risk communication. *Risk Analysis* 6(3):275–281.

Kemp, R. (1985). Planning, public hearings, and the politics of discourse. In *Critical Theory and Public Life*, ed. J. Forester. MIT Press, Cambridge, MA.

Kent. T. J, Jr. (1964). *The Urban General Plan.* Chandler Publishing Company, San Francisco.

Kingdon, J. (1984). *Agendas, Alternatives, and Public Policies.* Little, Brown, Boston.

Kunreuther, H. (1998). A program for reducing disaster losses through insurance. In *Paying the Price: The Status and Role of Insurance against Natural Disasters in the United States*, eds. H. Kunreuther and R. Roths. Joseph Henry Press, Washington, DC.

Kunreuther, H. (2006). Has the time come for comprehensive natural disaster insurance? In *On Risk and Disaster: Lessons from Hurricane Katrina*, Eds. R. J. Daniels, D. F. Kettl, and H. Kunreuther. University of Pennsylvania Press, Philadelphia.

McHarg, I. L. (1969). *Design with Nature.* John Wiley and Sons, New York.

Mileti, D. (1999). *Disasters by Design: A Reassessment of Natural Hazards in the United States.* Joseph Henry Press, Washington, DC.

Multihazard Mitigation Council, National Institute of Building Sciences (2005). *Natural Hazard Mitigation Saves: An Independent Study to Assess the Future Savings from Mitigation Activities. Vol. 1, Findings, Conclusions, and Recommendations.* National Institute of Building Sciences, Washington, DC.

National Research Council (2006). *Drawing Louisiana's New Map: Addressing Land Loss in Coastal Louisiana.* The National Academies Press, Washington, DC.

Nolon, J. and P. Salkin (2006). *Land Use in a Nutshell.* Thompson/West, Saint Paul, MN.

Olshansky R. and J. Kartez. (1998). Managing land use to build resilience. In *Cooperating with Nature: Confronting Land-Use Planning for Sustainable Communities* ed. R. Burby. Joseph Henry Press, Washington, DC.

Olson, R. S. (2000). Toward a politics of disaster: losses, values, agendas, and blame. *International Journal of Mass Emergencies and Disasters* 18(2):265–287.

Peacock, W. G., B. H. Morrow, and H. Gladwin (2000). *Hurricane Andrew: Ethnicity, Gender and the Sociology of Disasters.* Florida International University, Miami.

Platt, R. (1999). *Disasters and Democracy.* Island Press, Washington, DC.

Prater, C. and M. Lindell (2000). Politics of hazard mitigation. *Natural Hazards Review* 1(2):73–82.

Raiffa, H. (1982). *The Art and Science of Negotiation.* Belknap Press, Cambridge, MA.

Rubin, C. and I. R. Renda-Tenali (2001). *Disaster Timeline: Selected Events and Outcomes (1965–2000).* Claire Rubin and Associates, Arlington, VA.

Schwab, J., K. Topping, C. Eadie, R. Deyle, and R. Smith (1998). *Planning for Post-Disaster Recovery and Reconstruction.* PAS Report 483/484. American Planning Association Press, Chicago.

Smith, G. (2004). *Holistic Disaster Recovery: Creating a Sustainable Future.* Federal Emergency Management Agency. Emergency Management Institute, Higher Education Project.

Smith, G. and D. Wenger (2006). Sustainable disaster recovery: operationalizing and existing agenda. In *Handbook of Disaster Research*, eds. H. Rodriguez, E. Quarantelli, and R. Dynes. Springer, New York.

So, F. and J. Getzels (1988). *The Practice of Local Government Planning.* International City Managers Association, Washington, DC.
United States Department of Homeland Security (2007). *National Response Framework.* Draft. September 10. Washington, DC.

Chapter 10

Disaster-Resilient Communities

A New Hazards Risk Management Framework

Objectives

The study of this chapter will enable you to:

1. Link the concepts discussed throughout the text to hazard mitigation policy making.
2. Understand the strengths and weaknesses of existing hazard mitigation policies and programs, including the need for an improved hazards risk management policy framework.
3. Understand the proposed hazards risk management policy framework.

Key Terms

Community rating system
Federal Response Plan
Flood damage prevention ordinance

Flood Insurance Act of 1968
Flood Insurance Rate Map
Four phases of emergency management
Hazards risk management policy framework
Losses avoided study
Multihazard Mitigation Council
National Flood Insurance Program
National Mitigation Strategy
National response framework
National Response Plan
No Adverse Impact
Project Impact

Introduction

A comprehensive hazards risk management policy framework does not currently exist in the United States (Government Accountability Office 2007; Mileti 1999), nor are programs and policies guided by a philosophy or doctrine.* The means used to address natural hazards are more accurately characterized as a collection of disjointed policies directed by various agencies without a coordinative mechanism to guide collective action. Several key hazard mitigation policies will be discussed next, including their potential to help local communities become more disaster resilient. The chapter will conclude with a proposed hazards risk management policy framework and a series of recommendations to improve the nation's commitment to taking action in a more systemic manner based on a sound understanding of hazard vulnerability.

Current Hazard Mitigation Policies

The Stafford Act was developed to improve the level of coordination across federal agencies responsible for the administration of emergency management–related programs and policies, including many of those associated with hazard mitigation. The Federal Response Plan was intended to provide the operational guidance for the programs codified in the Stafford Act. This document, which focused on federal roles and responsibilities, was amended and renamed the National Response Plan, following the incorporation of state and local government responsibilities. After 9/11 and Hurricane Katrina the document was further modified and named the

* See the Emergency Management Institute Higher Education Project website for a proposed hazards management doctrine (http://training.fema.gov/EMIWeb/edu/).

National Response Framework. Each plan emphasizes response, broadly defined. As noted in the National Response Framework, "The term 'response' as used in this Framework includes immediate actions to save lives, protect property and meet basic human needs." The document goes on to state that "Response does not include prevention, protection or long-term recovery and restoration activities needed by communities to rebuild their way of life." (United States Department of Homeland Security 2007: 1). Hazard mitigation, a key element of the widely recognized four phases of emergency management (preparedness, response, mitigation, and recovery) is not addressed by the federal government through the National Response Framework or any other comprehensive policy framework.* While the Stafford Act and the Disaster Mitigation Act represent the primary federal legislation authorizing hazard mitigation, they are not linked to the array of other federal mitigation policies or programs nor other response and recovery activities.†

The Disaster Mitigation Act of 2000 represents a movement toward a more proactive set of activities guided by state and local hazard mitigation planning efforts rather than the reactive postdisaster grant and aid programs spelled out in the Stafford Act. The Disaster Mitigation Act has linked access to postdisaster funding to the development of a hazard mitigation plan. In concept, the connection between the development of a hazard mitigation strategy and federal assistance represents a step in the right direction. In practice, communities that have failed to develop hazard mitigation plans prior to a disaster have been allowed access to postdisaster assistance, while many plans created pre-event have failed to adequately address unsound land-use practices in known hazard areas or implement specific projects that target vulnerable structures identified in their hazard vulnerability analysis.

These problems can be traced, in part, to the limited emphasis placed on local capacity building. The creation of the Pre-Disaster Mitigation program, which has funded the development of most hazard mitigation plans and a number of hazard mitigation projects across the country and is indicative of a more progressive strategy, does not place an adequate emphasis on local capacity building. The federal program, Project Impact, which sought to develop predisaster public–private partnerships and build local capacity to address hazard vulnerability, was discontinued and replaced with the Pre-Disaster Mitigation Program (Birkland 2006). The creation of local hazard mitigation plans offers significant promise as a process through which a grass-roots effort can emerge. However, for this to prove successful across communities of varying capabilities, a greater commitment is required among

* FEMA and the Department of Homeland Security have placed greater emphasis on preparedness, response, and recovery, subsuming hazard mitigation under the concept of preparedness (United States Department of Homeland Security 2007).

† Additional federal legislation exists that emphasizes hazard-specific mitigation-related activities, including the Flood Insurance Act of 1968, the National Earthquake Hazards Reduction Act of 1977, and the National Windstorm Impact Reduction Act of 2004. The repeated adoption of narrowly focused legislation, rather than a broader hazard mitigation policy framework, has tended to follow major disasters (see Rubin and Renda-Tenali 2001).

federal and state agencies to institute targeted training programs while educating stakeholders about the merits of embracing and institutionalizing this approach.

The National Flood Insurance Program (NFIP), established by the Flood Insurance Act of 1968, represents the nation's most comprehensive program to reduce hazard-related losses in the United States.* The NFIP requires participating communities to adopt a local flood damage prevention ordinance that stipulates the type and location of development that can occur in a regulated floodplain. The Flood Insurance Rate Map (FIRM) designates estimated floodplain boundaries, approximates expected flood depths associated with differing flood return periods, and is used to calculate flood insurance rates. By regulating development that occurs in the floodplain, members of the community are able to purchase flood insurance. Some communities choose to enact additional measures to reduce flood hazard–related losses. Among the most effective programs is the Community Rating System. The Community Rating System (CRS) is an incentive-based program that results in reduced flood insurance rates for those policyholders living in a CRS community. Points are given for various hazard mitigation activities, and as they reach cumulative thresholds, flood insurance rates are reduced in accordance with assigned percentages (Figure 10.1).

Even though the NFIP represents the most advanced program available to confront natural hazards, it has had the unintended effect of encouraging development in flood hazard areas and increasing community exposure (Burby and French 1981; Burby et al. 1985). This has contributed to flood-related losses, particularly when flood events exceed the NFIP regulatory standards adopted by local governments. In an attempt to reduce incentivizing development in flood hazard areas and reduce

Credit Points	CRS Class	Premium Reduction (SFHA)
4,500+	1	45%
4000–4499	2	40%
3500–3999	3	35%
3000–3499	4	30%
2500–2999	5	25%
2000–2499	6	20%
1500–1999	7	15%
1000–1499	8	10%
500–999	9	5%
0–499	10	0

Figure 10.1 Community rating system categories and affiliated flood insurance rate reductions.

* Flooding represents the most significant natural hazard in the United States. Water related damages account for over 75 percent of federal disaster declarations and over 6 billion dollars in annual losses (Association of State Floodplain Managers 2003).

the escalating trend in flood-related damages, the Association of State Floodplain Managers have developed the program No Adverse Impact (NAI). This initiative is designed to encourage the local adoption of a more comprehensive flood loss reduction strategy that includes the adoption of higher NFIP standards, joining the CRS program, the practice of multiobjective management, adopting the principles of sustainability as a guiding set of principles, and the implementation of broad floodplain regulations (i.e., land-use and building codes, storm water management, protection of natural systems, and planning) (Association of State Floodplain Managers 2003). While the NAI suggests an approach that includes key themes described in the previous chapter, they are not linked to a national, codified strategy associated with specific incentives and penalties that increase the likelihood of compliance. The next section proposes a new risk management policy framework that requires communities to implement a more rigorous hazard mitigation strategy while helping them to develop the local capacity necessary to implement and sustain it.

A New Hazards Risk Management Policy Framework

The analysis of hazards provides the factual basis for action manifested in plans, policies, programs, and actions taken by individuals, groups, and organizations. The effective transfer of knowledge to action through hazards analysis and planning must recognize the existing social, political, and organizational conditions that shape outcomes. The characterization of hazards, the likelihood of their occurrence, the exposure of people and the built environment, the tools used to model their impacts, the methods chosen to communicate risk, and actions taken to reduce the impact of hazards are based on human constructs and decision-making processes (see Chapter 2). The policy framework described next integrates the concepts discussed throughout the text and provides an improved hazards policy that captures the lessons learned across hazard risk management strategies employed in the United States (see Figure 10.2).

We suggest a model that includes eight important elements: (1) develop and implement dynamic risk assessment tools that address current as well as expected future exposure and losses; (2) use this information to develop plans, programs, and policies that reflect this reality, including a direct link to land use and human settlement patterns; (3) increase the emphasis on community-level capacity building and self-reliance; (4) modify those policies that incentivize unsound pre- and postevent development relative to hazards and limit federal assistance to those that fail to comply; (5) hold communities accountable for their actions once new policies are created and the capacity necessary to effectively implement them is developed; (6) couple an all-hazards insurance program to an incentive-based hazard mitigation strategy; (7) document the merits of this approach at the community level through the quantitative and qualitative study of hazard loss avoidance, and incorporate the findings

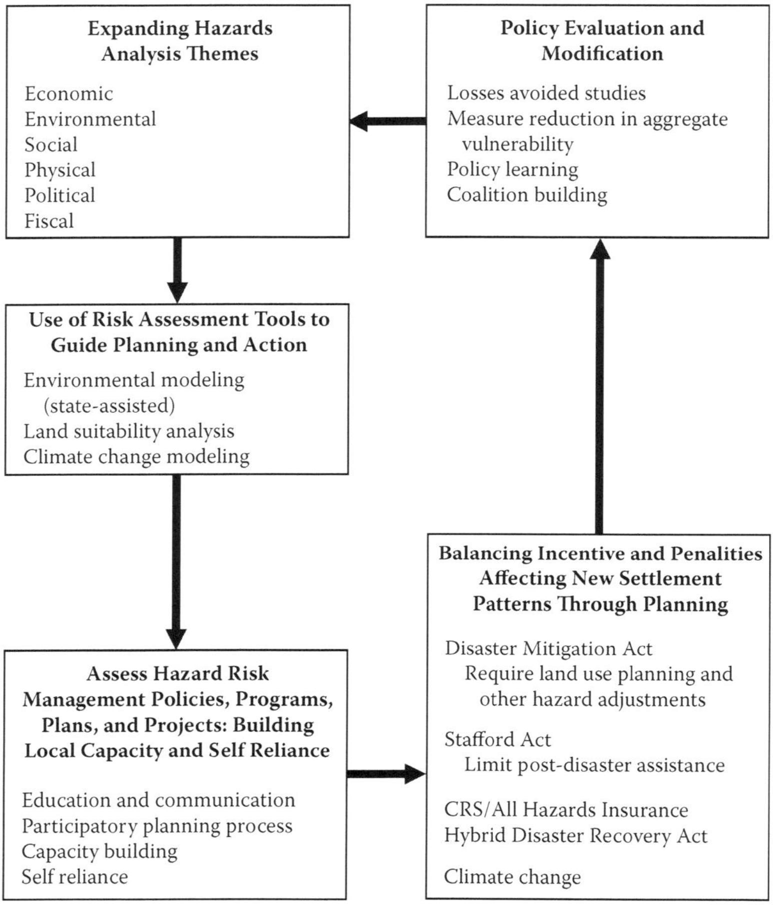

Figure 10.2 Hazards risk management policy framework.

into the local hazard mitigation planning process; and (8) build a broad coalition of support for the creation and maintenance of disaster-resilient communities.

Expand Hazards Analysis to Include Sustainable Development and Disaster-Resilient Themes

The process of hazard analysis should incorporate factors that define sustainable and disaster-resilient communities, including economic, environmental, and social conditions as well as measures of local capacity and self-reliance. Expanding the definition of a hazards analysis requires applying the methods discussed throughout the text, including the incorporation of a community's fiscal, technical, administrative, legal, and political capability. A high level of local capability increases a community's ability to effectively confront hazards and their related impacts. It also reduces the need for outside assistance before and after disasters, making them more self-reliant, an important but often underemphasized part of sustainable and disaster-resilient communities. Substantial, yet measured and ongoing organizational and institutional changes reflected, for example, in human settlement pattern adjustments should be made over time based on the results of this broader assessment.

Use Risk Assessment Findings to Guide Land-Use Planning and Action: Assessing Current and Future Vulnerability

The use of risk assessment findings to shape future land use and human settlement patterns benefits from a procedural and spatially oriented approach common to planning practice. Gaining acceptance from those impacted by proposed changes in land-use policy is enhanced when they are involved early and often in decision-making processes as the Charlotte/Mecklenburg case study in Chapter 9 aptly demonstrates. Land suitability analysis provides a sound way to visually and analytically link sustainable development themes to geographically defined hazards and comparatively evaluate the appropriateness of various development scenarios. The ability to adopt this approach requires a greater commitment from state and federal agencies to assist communities with the data collection and analysis needed to effectively use these tools. It also requires building the local capapacity needed to sustain an ongoing hazards analysis process as part of a larger hazard mitigation and disaster-resilient community program. Building local capacity increases the likelihood of adopting a regular reanalysis of hazards that reflects both the implementation of a community's hazard mitigation plan and changes in hazard vulnerability associated with new development.

Assess Hazard Risk Management Policies, Programs, Plans, and Projects: Building Local Capacity and Self-Reliance

Hazards analysis requires assessing the ability and willingness of organizations to adopt policies, programs, and plans addressing hazards and their potential effects. The process is also intended to identify weaknesses and develop a strategy to address them. Increasing local capacity and self-reliance should be the long-term aim of this approach. Conducting an inventory of existing tools helps to frame how members of a community perceive risk, including its political salience relative to other competing policy agendas. The actions taken by groups can decrease or increase exposure to hazards. For example, economic development policies may create incentives for growth while inadvertently increasing hazard vulnerability by placing new development in known hazard areas. Environmental preservation programs may limit development in environmentally sensitive areas that are also prone to hazards. In order to capture the collective impact of policies, programs, and plans, they should be assessed across federal, state, and local government agencies, businesses, nonprofits, financial and lending institutions, the insurance industry, and regional planning organizations. All play a role in shaping a community's vulnerability to natural hazards. A significant weakness in the current U.S. system is that it fails to adequately incorporate these interests into the hazard mitigation planning process.

Balancing Incentives and Penalties Affecting Human Settlement Pattern Adjustments: A Critical Look at Existing Hazard Mitigation Programs

A number of policies, programs, and plans exist that address hazards risk management. Yet their effectiveness has been significantly limited due to three primary reasons: (1) Policies have been developed in isolation from one another without a broader framework or plan to guide collective action focused on reducing our nation's vulnerability to natural hazards and disasters. (2) The formulation of hazard mitigation policy has not been coupled with a sincere federal commitment to build capacity at the community level to implement hazard mitigation strategies. (3) Existing hazards risk management policy has not effectively linked reducing future hazards losses to existing and future land-use activities. As described in Chapter 9, the practice of planning is uniquely situated to play a lead role in addressing these critical weaknesses. Planning is process oriented, participatory, and coordinative in nature. Building local capacity through participatory planning activities is a widely practiced skill among professional planners, while land-use planning is the fundamental domain of the planner.

The analysis of key hazard mitigation policies are described next, including how they can be more effectively integrated into a comprehensive policy framework. The Stafford Act, the Disaster Mitigation Act, and the National Flood Insurance Program include elements from which a substantially improved approach can emerge. We propose that the framework could form the basis of a resurrected National Mitigation Strategy. The original National Mitigation Strategy outlined steps taken by the federal government to support community-based hazard mitigation (FEMA 1995). At present there is no such federal plan. Neither the Stafford Act nor the Disaster Mitigation Act addresses the means to coordinate hazard mitigation policies across programs. The Stafford Act emphasizes the programmatic requirements associated with the administration of grant-in-aid programs, not the role of planning. The Disaster Mitigation Act emphasizes planning, narrowly defined, rather than codifying the process necessary to integrate programs and policies or tackle the fundamentally important question of land use and hazard vulnerability. FEMA is the lead federal agency for both programs, yet has been reluctant to address land use and its role in hazard mitigation policy. This is not surprising, because the agency does not have a federal mandate to actively pursue the often contentious issues associated with federal involvement in local land-use planning.

In order to confront these obstacles the Stafford Act and Disaster Mitigation Act should be amended to require states and communities to adjust human settlement patterns based on the findings of a more rigorous hazards analysis. The failure to shift land-use patterns based on this information would result in limited or no federal disaster relief following a presidentially declared disaster.* The Disaster Mitigation Act, which mandates the development of hazard mitigation plans, underemphasizes land use and focuses instead on the identification of hazard mitigation projects. These projects tend to address structures that are at risk due to poor land-use decisions made in the past. Thus plans become retrospective rather than forward looking. The potential power of plan making is grounded in its future orientation, and should be used for this purpose.

The Disaster Mitigation Act does not clearly articulate, nor require action based on the results of the hazards analysis. Instead, proposed policies and projects may or may not be closely aligned with identified community vulnerabilities or adjacent jurisdictions that share a common regional threat. Improving the link between hazards analysis and mitigation strategies should become more clearly codified. First, communities should be required to describe the nexus between the results of the hazards analysis and identified actions and be held accountable for taking the steps necessary to reduce future expected losses. Second, communities should be required to measure the losses avoided over time as recommended projects and policies are completed or adopted. The ability to quantitatively measure their impact provides benchmarks for success and tangible (i.e., monetary) benefits to those who may question the necessity or efficacy of these activities. Losses avoided measures should include assessing the reduction in aggregate vulnerability across a geographic area or region. Aggregate vulnerability can be defined as the level of exposure and risk that accounts for the adoption of hazard mitigation policies, the implementation of hazard mitigation projects, and continued development in known hazard areas. This approach is a more meaningful measure of the efficacy of a comprehensive hazards risk management program and can be used to measure multijurisdictional efforts to address regional hazards.

In order to address the weaknesses identified throughout this chapter, we propose a new hazards risk management framework that emphasizes the adoption of an all-hazards "CRS" program coupled with an all-hazards insurance model. The

* The Disaster Mitigation Act stipulates that communities are not eligible for postdisaster hazard mitigation funds unless they have developed a federally approved hazard mitigation plan. This requirement is sometimes ignored following disasters. Instead, communities are allowed to develop the plan postdisaster in order to gain access to Hazard Mitigation Grant Program funds. Developing a plan in the aftermath of a disaster is often problematic considering the number of competing agendas, including the desire to return to normal as soon as possible, rather than taking the time required to develop a plan based on a deliberative, participatory process.

implementation of this program can be achieved, in part, through a new approach to all-hazards mitigation planning. FEMA is currently in the process of finalizing programmatic rules linking CRS and Disaster Mitigation Act planning requirements, the consolidation of mitigation grant programs, and developing a strategy to update their floodplain mapping program include a greater emphasis on the connectivity between map products, risk assessment, loss estimation tools, mitigation planning, and risk communication (FEMA 2007).

We propose taking this integrative approach a step further, linking hazard mitigation planning activities to tangible benefits, namely reduced all-hazards insurance rates for those residing in communities that have developed plans that meet the new, more stringent requirements. Specific improvements include (1) confronting human settlement pattern adjustments in hazard-prone areas, (2) establishing a more direct correlation between the findings of the hazards analysis and the mitigation strategies adopted and implemented, (3) developing the means to assist local communities with building capacity and self-reliance, (4) measuring losses avoided, and (5) using this information to revise and update local hazard mitigation plans and recalibrate insurance rates as the exposure to the damaging impacts of hazards is reduced.*

The development of an all-hazards insurance program offers promise and has become an increasing topic of conversation among policymakers following Hurricane Katrina. In the United States, natural hazards insurance programs are uncoordinated, they tend to be disaster specific, premiums do not always reflect actual risk, nor are adequate economic incentives provided to stimulate the adoption of hazard mitigation measures. Kunreuther notes that an all-hazards insurance approach offers two key benefits: a reduction in coverage uncertainties associated with hazard-related impacts (e.g., the wind versus water debate following hurricanes) and an opportunity to spread risk, thereby reducing the likelihood that insurance payouts following a major disaster will exceed insurance premiums collected from across the country (2006). The sustained success of a program of this type requires developing an insurance strategy where rates accurately reflect the risk in known hazard areas. Insurance can be a powerful hazard mitigation tool, particularly when it helps to foster the adoption of risk reduction measures while encouraging those directly impacted by hazards and disasters to pay for the costs associated with recovery when a disaster strikes (Kunreuther 1998: 214). Encouraging "good

* The CRS program has been criticized as too labor intensive and costly to implement. This is a real concern for communities with limited technical, fiscal, and administrative capabilities. A key part of an all-hazards CRS program would require a sincere commitment of federal and state-level technical assistance in the form of ongoing training and capacity building. Financial assistance, in the form of Pre-Disaster Mitigation funding as well as postdisaster Hazard Mitigation Grant Program funds should be available for associated start-up costs. Consideration should be given to the financial commitment of the insurance industry as sound pre-event planning should reduce future losses and associated insurance payouts following disasters.

behavior," including individual disaster resilience and self-reliance, requires ensuring that policyholders have access to information about the actions they can take to reduce risk and the monetary benefits of doing so.

Policy Evaluation and Modification: Assess Losses Avoided and Build Disaster-Resilient Communities

The efficacy of any program should be evaluated over time to ascertain whether it is meeting established goals and measureable objectives. Losses avoided studies measure the benefits of reduced expected losses associated with the implementation of hazard mitigation projects and policies. The evaluation and assessment of hazard mitigation is an evolving process and in need of a more systematic means to measure its effectiveness and apply the results to the larger hazards risk management policy framework. The Multihazard Mitigation Council, a congressionally appointed collection of hazard scholars tasked with assessing the benefits of hazard mitigation activities, found that hazard mitigation projects and larger programs like Project Impact provided a 4 to 1 benefit–cost ratio when comparing losses avoided versus program costs (Multihazard Mitigation Council 2005). While promising, the results of the Multihazard Mitigation Council study have not led to the widespread integration of hazard mitigation into the larger disaster preparedness or homeland security directives of the federal government. Crafting policy that will be accepted by multiple stakeholders who maintain differing perspectives requires taking the time necessary to collect and analyze relevant information and build coalitions of support. These coalitions should include technical experts, practicing professionals, and politicians in order to develop the institutional conditions needed to sustain the effort over time and across political election cycles at the federal, state, and local level.

While losses avoided studies tend to focus on reduced damages to physical structures, a more robust analysis should also measure how existing (and proposed) policies and projects impact economic, environmental, and social factors associated with the concept of sustainability and resilience. The use of loss estimation models should also account for future development and measure changes in aggregate vulnerability to natural hazards. This requires assessing how the collective application of hazard mitigation strategies (e.g., land use, building codes, relocation of at-risk structures) and projected development patterns affect community and regional vulnerability. For example, a community may adopt a more stringent building code (which requires new construction to be built to a higher standard) and relocate at-risk structures while they place new development in harm's way.

Critical Thinking: Critique the proposed framework, offering specific recommendations for improvement based on the information presented in the text or other reading material.

Recommendations for Action

The recommendations described next are intended to frame the continued debate calling for a coordinated hazard risk management strategy. We have proposed a new framework, achieved primarily through the expansion of existing policies and programs. In order for the framework to function effectively, several important procedural items should be addressed. They include the need to assess the efficacy of local hazard mitigation planning, particularly their role in shaping land-use decisions and settlement patterns; the need to develop an enhanced training program targeting local capacity-building strategies; the provision of tangible benefits to those communities that do the right thing, and withholding access to these benefits if they do not; and increasing the involvement of professional land-use planners in the hazards management framework, including their role in facilitating social change and policy learning.

Conduct National Assessment of Local Hazard Mitigation Plans

The passage and implementation of the Disaster Mitigation Act has resulted in the development and approval of local and state plans across the country.* Have they made communities safer? Who was involved in their creation, and what role has this had on plan quality? What are the key characteristics of successful hazard mitigation planning processes? Which techniques have proven most successful in reducing hazard vulnerability? At this point we simply do not know the answers to these basic questions. Godschalk et al. (1999) assessed the quality of state hazard mitigation plans prior to the adoption of the Disaster Mitigation Act. They found that the quality of the plans varied widely, and their implementation strategies were weak. This study predated the development of more rigorous planning requirements. Have the new rules promulgated in the Disaster Mitigation Act proven effective? A multimethod approach is suggested that includes a quantitative analysis of state and local plan content, emphasizing factors that measure (1) the quality of the hazards analysis, capability assessment, mitigation strategy, and implementation procedures (plan evaluation, monitoring, and updates); (2) the impact of state planning on community-level planning; (3) the linkage between the findings of the hazards analysis and chosen hazard mitigation actions; (4) the degree to which land-use adjustments were made based on the results of the background studies; and (5) the commitment of states and local governments to measure losses avoided over time.

---

* FEMA approved all state and most local government's hazard mitigation plans by 2005. They are in various stages of the implementation process. Many communities and states are updating their plans. This provides a rich source of information regarding their content and the degree to which the plans have resulted in a reduced level of hazard vulnerability.

A qualitative review of select plans should be used to gather in-depth information on the planning process, including (1) the level of involvement among stakeholder groups and their impact on decision making; (2) the degree to which land-use planners were involved; (3) the type and efficacy of participatory and plan-making tools used; (4) variations in state and community-level capability and its impact on plan content and implementation; (5) the use of spatial analysis tools, including modeling and GIS; and (6) the role of the hazard mitigation planning process in conveying risk. The findings should be used to address limitations in DMA plans through a training and capacity-building agenda.

Place a Greater Emphasis on Land-Use Decisions and Human Settlement Patterns

Among the greatest weaknesses of existing hazard mitigation plans is their failure to confront development in known hazard areas. This is due in large part to perverted hazards policies predating the passage of the Disaster Mitigation Act that encourage, rather than discourage communities from making choices that reduce exposure and risk. Addressing this reality will not be easy and requires a combination of incentives and penalties. It also necessitates developing a unified strategy that reduces or eliminates counterproductive policies. Otherwise, we will continue to repeat the mistakes of the past, rather than embracing a forward-looking agenda emphasizing the concepts of disaster resilience and sustainable communities.

Establish a Robust Training and Capacity-Building Approach

The adoption and routine update of hazard mitigation plans require an improved, more systematic means to train local, state, and federal officials as well as other stakeholders that are often excluded from the process of hazards risk management. An adequate training regimen must emphasize the need for the direct involvement of local land-use planners throughout the hazard mitigation planning process. Kartez and Faupel identified the lack of coordination among land-use planners and emergency managers in a national assessment they conducted in 1994. It remains an ongoing problem and is particularly troubling when discussed in the context of hazard mitigation planning.

It is incumbent on planners and emergency managers to lead efforts to develop an improved training approach that emphasizes the technical aspects of plan making and capacity building. Hazards analysis techniques should emphasize a locally derived approach reflecting local capabilities, vulnerabilities, and solutions. At the same time, the hazard mitigation strategy must meet more stringent standards than those in place today. In order to achieve this aim, local officials must be provided improved support from FEMA. Greater attention should be placed on a training agenda focused on building local capacity and self-reliance, an important, but underemphasized component of a disaster-resilient, sustainable community.

Provide Benefits to Communities that Develop Comprehensive Plans

The Disaster Mitigation Act stipulates that states and local governments must develop and adopt a hazard mitigation plan in order to remain eligible for pre- and postdisaster hazard mitigation funding. Consideration should be given to eliminating all forms of federal assistance should communities fail to adopt a hazard mitigation plan. The bulk of federal assistance following most federally declared disasters is associated with debris cleanup and the repair of damaged infrastructure. These funds should be withheld postdisaster if a community fails to adopt an approved plan. On the other hand, communities that adopt higher standards, including those that address land-use relative to hazards should be eligible for additional types of assistance, including reduced all-hazards insurance premiums described in the proposed hazards risk management framework.

States are eligible to receive an increased level of Hazard Mitigation Grant Program funds postdisaster if they develop an "enhanced plan." Proactive local governments should be similarly rewarded. A percentage of pre- and postdisaster assistance should emphasize training and capacity building. This would avoid the rich get richer syndrome where larger, wealthier communities are able to meet higher standards due to their preexisting financial, technical, and administrative capabilities. The federal government should utilize community officials that have developing plans that meet higher standards to train others as part of a comprehensive outreach effort.

Comprehensive Hazards Risk Management Benefits from the Involvement of Professional Land-Use Planners

Professional land-use planners possess an array of tools, methods, and processes that have been refined over a century of codified practice (Hall 1988). These tools, while directly applicable to the challenges facing communities prone to natural hazards, have been applied sporadically as part of a comprehensive hazards management strategy (Burby 1998). Chapter 9 described both plan making and the procedural nature of planning. These activities are the professional domain of land-use planners. This realization is not intended to imply that planners are the only ones capable of leading planning activities. Rather, their involvement as technical experts and facilitators should not be underutilized. Many land-use planners mistakenly believe it is the responsibility of emergency managers to develop hazard mitigation plans, or they fail to recognize the connection between hazard resilience and sustainable development, of which the latter is widely acknowledged as a key aim of planning practice (Berke et. al. 2006). In order to capitalize on this underutilized and fundamentally important resource base, an extensive outreach campaign should be implemented by the American Planning Association, FEMA, the International Association of Emergency Managers, the National Emergency Management Association, and others.

Facilitate the Use of Planners as Agents of Social Change and Policy Learning

The recommendations noted in this chapter reflect principal themes of planning practice and theory, including plan making, data collection and analysis, participatory planning and dispute resolution, and capacity building. Yet when discussed in the context of hazards, those individuals and organizations trained in these skills, namely land-use planners, are often excluded. Perhaps this provides some understanding as to why we have not done a better job of systematically addressing natural hazards in the United States. It also points out that planners have not proven as effective when confronting basic land-use challenges that are often driven by those in positions of power, namely what Logan and Molotch refer to as the "land-based rentier elite" (Logan and Molotch, 1987). Indeed, the Marxist view of planning defines the process as a means to maintain the capitalist city through the organization of private interests while reducing social conflict (Castells 1977). Planners and others supporting the concepts of sustainable development and disaster resilience must become advocates as Arnstein and Davidoff suggest, forming an effective coalition capable of advancing complimentary agendas.

Social learning theorists assume that lessons are translated into policy adaptations based on experience and the formulation of coalitions advocating change (Sabatier and Jenkins-Smith 1993). These concepts have been applied to hazard mitigation (Olson 2000) and disaster recovery (Birkland 1997, 2006; Smith and Wenger 2006). Birkland has emphasized the potential for disasters to serve as "focusing events" that can elicit change or highlight the failings of the policy system as Hurricane Katrina so dramatically demonstrated (2006). In order for planners to be effective change agents, it is important to understand both the impact of the preexisting conditions found in an area subject to natural hazards and disasters and also the triggers that may be elicited when an event occurs. Armed with this information, planners can serve several key roles—collecting and synthesizing information, displaying and analyzing spatial data, facilitating dialogue among stakeholders, and creating action-oriented recommendations as part of the plan-making process.

Include Hazards Analysis in Planning for Climate Change

The global implications of hazards have become increasingly salient as scientific consensus emerges regarding climate change. The realization that we are altering the planet in a way that could drastically affect our way of life, our economies, and future human settlement patterns brings us back to the basic tenets of sustainable development described in Chapter 9. The findings of the Intergovernmental Panel on Climate Change describe a world in which we can expect significant changes in ecosystems and biodiversity, the spread of infectious disease, increases in extreme heat and drought, and abnormally heavy precipitation in other locations (Intergovernmental Panel on Climate Change 2007). In coastal areas, where over 50% of the United States population currently resides, we can expect a rise in sea levels and an increase in tropical cyclone activity, including those of greater intensity. This will result in greater damages due to coastal erosion, flooding, and high winds in areas experiencing rapid development.

Globally scaled models display a wide range of possible impacts. The analysis of these hazards merits continued research in order to gain a greater understanding of the influence of multiple variables both alone and in combination with others. This will lead to an increased level of accuracy, both spatially and temporally. That is, to what extent will sea levels rise and at what speed? How can we predict with greater accuracy the onset of drought or increases in rainfall that could be used to develop more sustainable land-use practices? Can the threat of slow-onset, potentially catastrophic hazards associated with climate change facilitate the systematic changes needed to confront them?

The current U.S. system of hazards risk management, which tends to focus on natural events that are well understood in terms of their physical characteristics and their impacts on individuals, groups, organizations, and institutions, does not employ an integrated policy framework to address them. How then do we expect to engage in the type of long-term comprehensive planning necessary to address the exceedingly complex threats associated with climate change? The climate change dilemma provides us an unparalleled opportunity to improve the working relationship between those involved in hazards risk management and those tackling the broad challenges of climate change and sustainable development.

Slow-onset catastrophic hazards provide both the substantial threat and the time necessary to develop appropriately scaled plans and the coalition of support needed to implement identified recommendations. In reality, slow-onset hazards may offer even greater challenges as changes in exposure and vulnerability occur over time periods that exceed local election cycles. As a result, federal, state, and local officials may be resistant to adopt forward-looking policies that may prove politically controversial and whose benefits may not be accrued immediately. Advancing this agenda requires building a broad-based coalition that includes elected officials, local advocates, and organizations that derive their influence from grass-roots support and not just election results.

Critical Thinking: Are our communities capable of planning for slow-onset catastrophic events? Will slow onslaught of disasters foster the level of collaboration needed to tackle them?

Discussion Questions

Which policies do you think are most effective, those based on incentives or penalties?

Do you think the creation of an all-hazards insurance program is a good idea? Based on what you know about hazards, their geographic distribution, and their associated impacts, who do you think will support or oppose this concept?

Based on what you have learned in this text, how would you assess aggregate community vulnerability across the various hazards prevalent in the United States?

In addition to losses avoided studies, can you think of other ways to document the benefits of hazard mitigation?

How do you propose to encourage changes in existing land-use patterns relative to hazards?

Do you think a federal mandate is the most effective way to facilitate change at the local community level? Can you think of other alternatives?

Can you think of any recommendations in addition to those posed at the end of this chapter that merit attention? If so, what are they, and why do you think they are important?

Application

You Be the Policy Analyst

Based on your reading of the text and other sources of information, what other policies offer promise when addressing the challenge of reducing our vulnerability to natural hazards? Critique the proposed policy framework, identifying strengths and weaknesses while offering specific improvements. How do the implications of climate change affect your perspective?

References

Arnstein, S. R. (1969). A ladder of citizen participation. *Journal of the American Institute of Planners*. 35:216–224.

Association of State Floodplain Managers (2003). *No Adverse Impact: A Toolkit for Common Sense Floodplain Management*. Association of State Floodplain Managers, Madison, WI.

Berke, P., D. Godschalk, E. Kaiser, and D. Rodriguez (2006). *Urban Land Use Planning*, 5th Edition. University of Illinois Press, Champaign.

Birkland, T. A. (1997). *After Disaster: Agenda Setting, Public Policy and Focusing Events.* Georgetown University Press, Washington, DC.

Birkland, T. A. (2006). *Lessons of Disaster: Policy Change after Catastrophic Events.* Georgetown University Press, Washington, DC.

Burby, R. (1998). *Cooperating with Nature: Confronting Natural Hazards with Land-Use Planning for Sustainable Communities.* Joseph Henry Press, Washington, DC.

Burby, R. and S. French (1981). Coping with floods: the land use management paradox. *Journal of the American Planning Association* 47(July):289–300.

Burby, R., S. French, B. Cigler, E. Kaiser, D. Moreau, and B. Stiftel (1985). *Flood Plain Land Use Management: A National Assessment.* Westview Press, Boulder, CO.

Castells, M. (1977). *The Urban Question: A Marxist Approach.* Edward Arnold, London.

FEMA (1995). *National Mitigation Strategy.* Federal Emergency Management Agency, Washington, DC.

FEMA (2007). Higher Education Project. Federal Emergency Management Agency, Washington, DC.

Godschalk, D., T. Beatley, P. Berke, D. Brower, and E. Kaiser (1999). *Natural Hazard Mitigation: Recasting Disaster Policy and Planning.* Island Press, Washington, DC.

Government Accountability Office (2007). *Report to the Ranking Member, Committee on Financial Services, House of Representatives. Natural Hazard Mitigation: Various Mitigation Efforts Exist, but Federal Efforts do not Provide a Comprehensive Strategic Framework.* GA0–07–403, United States Government Accountability Office, Washington, DC.

Hall, P. (1988). *Cities of Tomorrow.* Basil Blackwell, Ltd., Oxford, England.

Intergovernmental Panel on Climate Change (2007). Summary for policymakers. In *Climate Change 2007: Impacts, Adaptation and Vulnerability.* Contribution of Working Group II to the Fourth Assessment Report of the Intergovernmental Panel on Climate Change, eds. M. L. Parry, O. F. Canziani, J. P. Palutikof, P. J. van der Linden, and C. E. Hanson. Cambridge University Press, Cambridge, England.

Kartez, J. and C. Faupel (1994). Comprehensive Hazard Management and the Role of Cooperation between Local Planning Departments and Emergency Management Offices. Unpublished paper.

Kunreuther, H. (1998). A program for reducing disaster losses through insurance. *In Paying the Price: The Status and Role of Insurance against Natural Disasters in the United States,* eds. H. Kunreuther and R. Roths. Joseph Henry Press, Washington, DC.

Kunreuther, H. (2006). Has the time come for comprehensive natural disaster insurance? In *On Risk and Disaster: Lessons from Hurricane Katrina,* eds. R. J. Daniels, D. F. Kettl, and H. Kunreuther. University of Pennsylvania Press, Philadelphia.

Logan, J. and Molotch, H. (1987). Urban Fortunes: The Political Economy of Place. University of California Press, Berkeley, CA.

Mileti, D. (1999). *Disasters by Design: A Reassessment of Natural Hazards in the United States.* Joseph Henry Press, Washington, DC.

Multihazard Mitigation Council, National Institute of Building Sciences (2005). *Natural Hazard Mitigation Saves: An Independent Study to Assess the Future Savings from Mitigation Activities. Vol. 1, Findings, Conclusions, and Recommendations.* National Institute of Building Sciences, Washington, DC.

Olson, R. S. (2000). Toward a politics of disaster: losses, values, agendas, and blame. *International Journal of Mass Emergencies and Disasters* 18(2):265–287.

Rubin, C. and I. R. Renda-Tenali (2001). *Disaster Timeline: Selected Events and Outcomes (1965–2000).* Claire Rubin and Associates, Arlington, VA.

Sabatier, P. and H. J. Henkins-Smith (1993). Policy Change and Learning: An advocacy approach. Westview Press.

Smith, G. and D. Wenger (2006). Sustainable disaster recovery: operationalizing and existing agenda. In *Handbook of Disaster Research,* eds. H. Rodriguez, E. Quarantelli, and R. Dynes. Springer, New York.

United States Department of Homeland Security (2007). *National Response Framework.* Draft. September 10. Washington, DC.

INDEX

W

Y

Z